Automated Fiber Placement:
Status, Challenges, and Evolution

Automated Fiber Placement: Status, Challenges, and Evolution

Ramy Harik and Alex Brasington

SAE INTERNATIONAL®

400 Commonwealth Drive
Warrendale, PA 15096-0001 USA
E-mail: CustomerService@sae.org
Phone: 877-606-7323 (inside USA and Canada)
            724-776-4970 (outside USA)
Fax: 724-776-0790

Library of Congress Catalog Number 2024945956
http://dx.doi.org/10.4271/9781468603064

ISBN-Print   978-1-4686-0305-7
ISBN-PDF   978-1-4686-0306-4
ISBN-epub   978-1-4686-0307-1

To purchase bulk quantities, please contact: SAE Customer Service

E-mail: CustomerService@sae.org
Phone: 877-606-7323 (inside USA and Canada)
            724-776-4970 (outside USA)
Fax: 724-776-0790

Visit the SAE International Bookstore at books.sae.org

Publisher
Sherry Dickinson Nigam

Product Manager
Amanda Zeidan

**Production and
Manufacturing Associate**
Michelle Silberman

This book is dedicated to the next generation of composite pioneers.
May it spark your own groundbreaking innovations!

Contents

Chapter 02 - AFP Process Parameters

Chapter 03 - AFP Defects

Chapter 04 - Process Planning

Foreword

Fiber-reinforced composites have revolutionized the advanced materials market in the last 50 years. The transformation has been particularly remarkable for the aerospace industry, where materials such as carbon fiber-reinforced polymer matrix composites (CFRPs) have enabled lightweight structures resistant to fatigue and corrosion. The use of CFRPs steadily grew in the 1970s, 1980s, and 1990s for medium-scale components such as light jet fuselages and large jet nacelles. There was then a step function change in the early 2000s with the introduction of prepreg automated fiber placement (AFP) for large fuselage and wing structures on the Boeing 787 and Airbus A350. AFP enabled less material scrap, directionally tailored properties, and an order of magnitude higher laydown rate. These changes enabled aircraft with approximately 50% composite content by weight, which significantly reduced fuel burn and improved longevity.

In the 20 years since, aerospace composite manufacturing has continued to push forward. New commercial aircraft (Airbus A220 wings, and Boeing 777X wings) and defense aircraft (Sikorsky CH-53K, Northrop Grumman B-21, Bell V-280, Airbus H160M, and Shenyang J-16) have continued to embrace composites and accelerate the 787/A350 baseline technologies. Raw materials have gradually improved, particularly higher-strength carbon fibers and highly toughened epoxy matrices. Meanwhile, prepreg AFP has roughly doubled in speed in the last 20 years through machine improvements, better maintenance, program optimization, and an increased skills base. Resin infusion has also emerged as an alternate technique to form integrated structures. Finally, thermoplastic composites have seen increasing use for high-rate stamp forming of smaller parts.

Despite these advances in composite fabrication, the industry is now at a tipping point where another revolution is needed to support several new markets. The rise of advanced air mobility (AAM) platforms will require light, complex composite structures at high rates. For mass commercial transport, several versions of a "next single aisle" (NSA) or "next midmarket aircraft" (NMA) are expected to launch in the 2030s. These new aircraft are anticipated to combine the lightweight architecture of the 787 and A350 with the high fabrication rates of the 737 and A320. Sustainability initiatives will also push these aircraft to use alternate fuel sources such as batteries (for AAM) and potentially hydrogen (for NSA/NMA). These fuel sources demand lightweighting and the use of composite battery cases and hydrogen tanks. In the defense sphere, collaborative combat aircraft (CCA) will require light complex structures similar to the AAM market. Finally, the drive for hypersonic missiles and aircraft will require more use of high-temperature ceramic matrix composites (CMCs).

Five major composite technology platforms are being pursued to provide lightweight, integrated, low-cost structures for these new markets. The first technology is thermoset prepreg, which is the baseline technology for most applications due to its current widespread use. Prepreg AFP is set to further evolve in the near term through collaborations between material suppliers, equipment manufacturers, and part producers. More precise heating, such as laser heating, is now being implemented for improved tack control. Actively fed "servo creels" are promoting more uniform tension and laydown accuracy. In-line inspection using on-head optics is reducing the downtime associated with visual inspection. Machine advances are enabling better tow steering/shearing, multiple heads working at once, and interchangeable heads for parallel processes such as three-dimensional (3D) printing. Finally, software advances are enabling better path optimization while addressing defects such as "laps and gaps." Beyond AFP, automation is extending into other processes such as the manipulation of fabric plies and subcomponents such as stringers. Once parts are laid up, new rapid cure and out-of-autoclave prepregs are set to enable lower cycle times and reduced energy use.

Finally, multifunctional material architectures are being investigated that would allow functions such as damage sensing, self-healing, thermal management, and shape morphing.

The second material platform is thermoset resin infusion, which provides advantages such as lower cost material, more integrated parts, and matched mold tooling control. Dry fiber placement (DFP) is evolving to match current AFP technology, which would enable better material usage and property optimization versus standard fabrics. DFP presents new challenges with how to use binders for proper tack and fiber control. Automated techniques are also being developed to produce narrow fabric structures such as stringers through continuous deposition and forming. Finally, braiding, stitching, and complex tooling schemes are being developed to allow for more integrated preforms before cure. Once a preform is ready, resin can then be infused through techniques such as resin transfer molding (RTM) and resin transfer infusion (RTI). Isothermal resins are being developed such that parts can be quickly cycled in and out of molds.

Thermoplastic continuous fiber composites represent the third major material platform. Thermoplastics provide advantages such as fast cycle times, welding ability, and better end-of-life recycling. Stamp forming is already used in the industry for small parts such as clips, and the industry is now researching how to stamp form larger structures such as frames. Thermoplastic AFP is emerging as a technique to both create blanks for stamp forming and to create large structures such as fuselage panels. Thermoplastic AFP presents challenges due to high working temperatures and difficulty achieving in-situ consolidation. One promising method for consolidation is to apply heat and vacuum pressure after laydown, while simultaneously "cofusing" other details such as stringers. Thermoplastic details can then be welded with several techniques (induction, resistance, and ultrasonic). More mechanical testing is still needed to validate thermoplastic materials and weld techniques, since these materials have not seen the service history of thermosets to date.

The fourth major material platform is CMCs. CMCs such as C/C, C/C-SiC, Ox/Ox, and SiC/SiC offer extremely high-temperature resistance compared to polymer matrix composites. However, these materials are traditionally very expensive, require long cycle times, and are not well suited for fabricating large complex structures. Several research and development (R&D) efforts have been initiated to address these shortcomings. One path is to improve the processes where dry fiber preforms are infused with resin and then pyrolyzed. The second path is to improve the use of precursor prepregs, which can be processed similar to thermoset prepregs before an additional pyrolysis step. Raw materials are also being optimized to reduce overall system cost.

The final material platform is the 3D printing of continuous fiber composites. While not suited for large acreage structures, 3D printing can enable the design of light, extremely complex shapes with minimal support tooling. Machines can now print dry fiber (for infusion with thermoset or ceramic matrices), thermoset prepreg, or thermoplastic prepreg. Creative topology-optimized designs can be built due to the fluidity of fabrication. One key drawback with 3D printing is the lack of data on mechanical performance, including the effects of print direction and defects.

Regardless of how composite details are built, breakthroughs are also needed for assembling them and reducing fastener weight. For thermoset prepreg structures, bonding technology is advancing through better adhesives and more automated surface preparation techniques (sanding, plasma, and laser). Infused parts can be co-infused initially or similarly bonded with adhesive later. Thermoplastic structures can be welded with a variety of techniques. CMCs can potentially be bonded, but there are challenges with using high-temperature adhesives. More data is still needed to certify structures with the same confidence as traditionally fastened parts for all of these joining techniques. In the meantime, fastening itself can still be optimized through more precise techniques such as full-scale determinate assembly (FSDA).

Finally, the composites industry must embrace the digital tools of "Industry 4.0" to enable new architectures at high rate and low cost. Advances in robotics are allowing the automation of nearly any repetitive task. Smaller, mobile robots capable of working side-by-side with humans are especially useful to reduce the need for expensive factory "monuments." Robot and machine efficiency is increasing with advanced analytics, especially when using machine learning (ML) and artificial intelligence (AI). Part inspection is rapidly evolving, since digital tools can quickly compare "as built" conditions to ideal conditions and track trends for process improvement. Finally, new software tools are enabling "digital twins" that connect physical parts to their digital counterparts. This enables better tracking of parts throughout fabrication, optimal day-by-day factory scheduling, and the ability to track how part defects might influence downstream processes. The goal is to produce parts as efficiently as possible while providing a lifelong digital pedigree for each one.

We stand at a critical juncture in the history of composites. New aerospace markets are producing unprecedented demand for lightweight integrated composite structures. NSA, CCA, and EVTOL industrial requirements will demand another order of magnitude rate acceleration relative to the 787/A350 reference. Smart manufacturing and advanced analytics coupled with much-improved automation platforms will enable AFP and DFP as primary routes to build these new structures at these much higher rates, while also enabling lower cost and improved quality diagnostics.

Aerospace composite advances should also trickle down to other industries such as automotive, wind energy, marine, construction, and sporting goods. This is an incredibly exciting time to work in composite manufacturing!

Sean Black, PhD, FIMechE, FRAeS **Stephen Pety, PhD**
Chief Technology Officer   Senior R&D Engineer
Spirit AeroSystems   Spirit AeroSystems

Acknowledgments

This book would not have been possible without the inspiration and motivation we have drawn from the neXt team members over the past decade. Your inquisitive minds, great team spirit, and passion for composites and manufacturing have continually pushed us to strive for excellence. To Dr. Harik's PhD students—Mazen Al Bazzan, Luis Bahamonde, Alex Brasington (the awesome co-author), Wout De Backer, Fadi El Kalach, Benjamin Francis, Victor Gadow, Matthew Godbold, Joshua Halbritter, Christopher Sacco, Clint Saidy, Philip Samaha, Roudy Wehbe, Kaishu Xia, and Ibrahim Yousif—your research, insights, and collaborations have profoundly shaped my understanding and the findings within the chapters of this book. To the master's students—Ahmed Mahmoud, Anis Baz Radwan, Benjamin Greenberg, Devon Clark, Drew Sander, Hossein Ahmed, Ibrahim Sarikaya, Jad Samaha, Max Kirkpatrick, Nishan Patel, Nitol Saha, Noah Swingle, Rowen Burney, Teddy Tarekegne, William Montgomery, and Yang Shi—your enthusiasm and fresh perspectives have been invaluable, and your grit to push forward is inspiring. And to the numerous undergraduate students who have been neXt UG researchers, your curiosity and eagerness to learn have consistently filled us with awe.

We extend our sincere gratitude to all the reviewers who graciously volunteered their time to review our chapters. Anis Baz Radwan, August Noevere, Matthew Godbold, Noah Swingle, Roudy Wehbe, and Waruna Seneviratne—your meticulous attention to detail made this book tenfold better. This book is in its first edition, marking the beginning of a journey. We eagerly anticipate revising it every few years, ensuring we continue to drive the exciting field of AFP forward.

We are deeply grateful to Dr. Brian Tatting for his thorough review of this book. His extensive and insightful feedback has been invaluable, and we have incorporated the majority of his excellent suggestions into this edition. We look forward to collaborating further on his more detailed comments for future editions.

We are very grateful to our industrial partners who provided us with many of the images! Special thanks to Ingersoll Machine Tools, ElectroImpact, Fives and Mikrosam. We have full intentions to knock on the door of all AFP equipment manufacturers in round two in a couple of years! Also grateful to SAMPE for allowing us to reuse some of our published work in the SAMPE/CAMX conferences and adapting it for the book.

We are indebted to Samar Mouawad for her talent in creating the stunning graphics that grace this book. Her artistry has not only enhanced the visual appeal but also deepened the understanding of the concepts presented. Reach out to us for any of the graphics, and please do use them in your classrooms!

We would like to express our appreciation to Sherry Nigam, our awesome publisher, for her unwavering support and much-needed patience. This is our third book! And there is no way this could have been done without your belief in us. Amanda Zeidan, your meticulous attention to detail and tireless efforts have been instrumental in bringing this book to the finish line in its current format! You are such a pleasure to work with and deal with! We would also like to acknowledge Linda and Bruce; when this project stalled, the memory of Linda's positivity and belief in us ignited the determination needed to keep moving forward.

Finally, we would like to thank our awesome families for always being there for us. You provide us with the strength needed to keep pushing forward!

Introduction

The dream of conquering space and building structures on the Moon, Mars, and beyond has fascinated the authors since childhood. This ambition holds immense potential, offering humanity an opportunity to expand beyond Earth, potentially mitigating territorial conflicts and fostering peace. While this dream may be beyond our current reach, advancements in innovation and technology, particularly in composite materials, are bringing it closer to reality. The authors believe that composite materials, with their exceptional strength-to-weight ratio, are the key to constructing sustainable structures in space. Lighter structures require less energy to operate, enabling the use of renewable sources like solar and wind power.

Among the various manufacturing techniques for composites, AFP stands out as the most promising. AFP's versatility allows for the creation of complex shapes and pushes the boundaries of composite design. Until new methods can replicate the material properties of continuous fibers, AFP offers the most viable path toward achieving our space construction ambitions.

The authors' fascination with AFP stems from its potential to accelerate the development of space exploration and unlock new horizons for humanity. By advocating for wider adoption and understanding of AFP, they hope to inspire others to join this exciting journey. This book aims to advance the field of AFP by promoting its wider adoption and highlighting its potential benefits across various industries, including automotive, mobility, and even medicine. It serves as both a research resource and an educational tool, presenting a decade of accumulated research in a clear and cohesive manner to engage readers with this complex technology.

The book offers a concise yet comprehensive exploration of AFP, incorporating various elements to facilitate a quick but deep understanding. By reading this book and watching the accompanying videos, users can rapidly progress from novices to knowledgeable practitioners in the field. With breakthroughs in materials, machinery, and configurations occurring constantly, this book provides a generalized overview to equip readers with essential information. It is designed as a starting point for graduate students and engineers entering the field—a "Welcome to AFP" guide that kickstarts their journey.

We are excited about the future of this book and envision it evolving with subsequent editions, incorporating the latest advancements and research in the dynamic world of AFP. This book is just the beginning; our hope is that it inspires readers to embark on a journey of discovery, unlocking new technologies and pushing the boundaries of what is possible with AFP.

Visit afpbook.com for images and other materials you can use in your publications.

List of Acronyms

2D - Two-Dimensional

3D - Three-Dimensional

AAM - Advanced Air Mobility

ABS - Acrylonitrile Butadiene Styrene

ACSIS - Automated Composite Structures Inspection System

AE - Acoustic Emission

AFP - Automated Fiber Placement

AHP - Analytic Hierarchy Process

AI - Artificial Intelligence

ATL - Automated Tape Laying

BO - Bayesian Optimization

CAD - Computer-Aided Design

CAPP - Computer-Aided Process Planning

CCA - Collaborative Combat Aircraft

CFRPs - Carbon-Fiber-Reinforced Polymers

CMCs - Ceramic Matrix Composites

CT - Computed Tomography

DFM - Design for Manufacturing

DFP - Dry Fiber Placement

DoD - Department of Defense

EI - ElectroImpact

FAA - Federal Aviation Administration

FFF - Fused Filament Fabrication

FOD - Foreign Object Debris

FPGAs - Field Programmable Gate Arrays

FSDA - Full-Scale Determinate Assembly

FW - Filament Winding

GPR - Gaussian Process Regression

GPUs - Graphical Processing Units

HGTs - Hot Gas Torches

ID - Identity

IM - Intermediate Modulus

IR - Infrared

ISAAC - Integrated Structural Assembly of Advanced Composites

LLO - Laminate-Level Optimization

ML - Machine Learning

NASA - National Aeronautics and Space Administration

NC - Numerical Code

NDT - Non-Destructive Testing

NMA - Next Midmarket Aircraft

NSA - Next Single Aisle

NURBS - Non-Uniform Rational Basis Spline

OOA - Out-of-Autoclave

PAN - Polyacrylonitrile

PLM - Product Lifecycle Management

PLO - Ply-Level Optimization

PPE - Personal Protection Equipment

QR - Quick-Response

R&D - Research and Development

RIPITx - Real Time in Process AFP Inspection

RTI - Resin Transfer Infusion

RTM - Resin Transfer Molding

STL - Stereolithography

UT - Ultrasonic Testing

VARTM - Vacuum-Assisted Resin Transfer Molding

VCP - VERICUT Composite Programming

WIP - Work in Progress

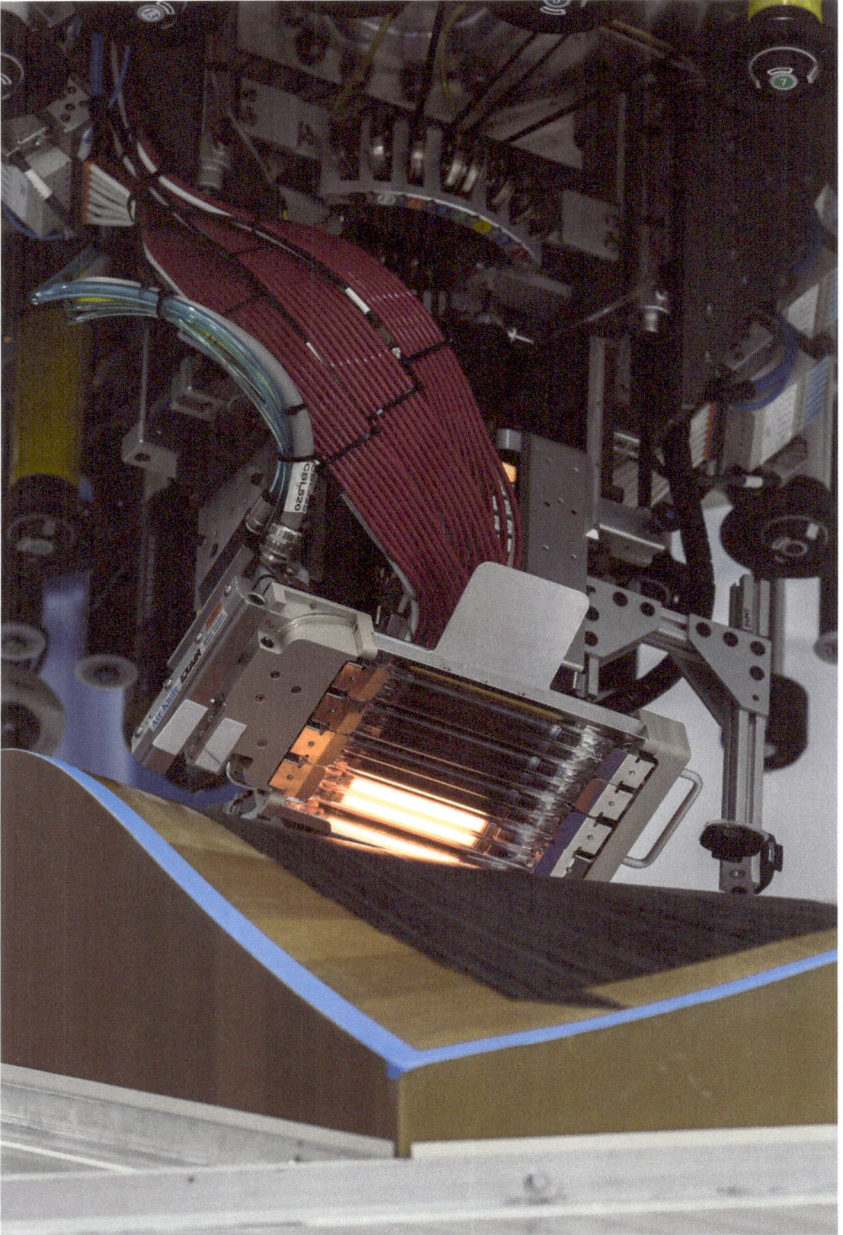

Electroimpact AFP machine at NASA Langley in Virginia.

Chapter 01

Introduction and History

Future sustainable structures require advanced materials and manufacturing techniques to adapt and integrate the ever-increasing throughput, quality, and repeatability demands. AFP and composite materials are the leading combination to achieve such a feat. This book presents the current status and challenges of AFP and identifies the required advancements to evolve AFP and composite materials. The combination of AFP and composite materials is ideal for promoting economical and sustainable development through various applications, such as space shelters, efficient aircraft, racing cars, and energy harvesting.

To successfully garner the benefits of AFP and composite materials, a solid foundation of the intricacies of both is required. This chapter will thoroughly define (1) composite materials, (2) composite manufacturing, and (3) the history of AFP, prior to expanding and developing further details of AFP in subsequent chapters. We will discuss the definition, material forms, and micromechanics of composite materials in the first section. Then, we will expand on different composite manufacturing techniques, detailing form-specific and generic ones. Finally, we will introduce a timeline of AFP history and major milestones, setting the landscape for upcoming chapters.

1.1.
Composite Materials

Composite materials are heterogeneous combinations of two or more constituents with different mechanical, physical, and/or structural properties. The fundamental concept is that a composite material shows superior properties for a specific engineering application compared to its individual constituents. Composites are not a new development and are quite common in nature. For example, materials such as wood and bones are composites and have been arguably around for millions of years.

There are many forms of composite materials, but typically, a composite contains a reinforcement component that provides stiffness and/or strength, and a matrix that surrounds the reinforcement component holds the composite in its final shape and provides load distribution to the reinforcement component. Reinforcement components are of various forms and shapes, from short and/or small particles to long continuous ones. The latter is usually the most desired type since it offers excellent load-carrying properties, often resulting in highly desirable strength-to-weight or stiffness-to-weight ratios. From the perspective of the manufacturing engineer, however, the former type, sporting short/small particles, is often considered more desirable as it significantly reduces the complexity of design and manufacturing.

Today, fiberglass is one of the most frequently used composite materials. In fiberglass, the reinforcement is made of continuous glass fibers, while the matrix is made from a resin system. In the aerospace industry, CFRPs are one of the most advanced engineering materials, showing highly desirable properties when designed and manufactured correctly. The high strength and high stiffness provided by the carbon fibers (the reinforcement component) offer unmatched properties. Thermoset polymers are most commonly used today as the matrix component; however, thermoplastic polymers are becoming more common as an alternative matrix system for CFRPs. Recyclability is an unresolved issue for many composite materials, and this switch toward thermo-plastics as matrix components might make composites even more appealing in the future due to their ability to be recycled.

Introduction to Composites

Duration: 23 minutes

Description: This video, part of "Composites A-Z: 30 Days of Composites," introduces composites by explaining their basic makeup, distinguishing between inhomogeneous and anisotropic materials. It covers the roles of reinforcement (fibers) and matrix materials, detailing how they work together to create strong, lightweight structures. The content lays a foundation for understanding composite behavior and performance in engineering applications.

Scan the QR code below to watch the video.

Numerous historical examples of composite products can be used to illustrate the fundamental concept and highlight the differences from traditional, non-composite engineering materials. In civil construction, an "adobe" (in Spanish) is a brick made from mud as the matrix material and straw as the reinforcement material. These bricks are famously used in a three-layer construction structure, including a first layer of adobe, followed by layers of mud and limestone. These homes were famous for this "composite" stone, enabling them to be one of the most comfortable dwellings as they provided cool temperatures in summer and warm temperatures in winter. This example highlights the fundamental concept that composites enable and unlock better properties than their individual elements.

Composite products are widespread across multiple industrial domains, and their usage has been on the rise since the 1960s. Although the aerospace industry has been at the forefront of utilizing and spreading the use of composites and composite products, numerous other industries are currently expanding the limits of composites' design and production. Composites were around in sporting goods long before they were used in aircraft since people were willing to pay huge amounts of money to get a slightly better golf club or tennis racket. In addition, boat hulls with high porosity have been around for quite a while since weight is not an issue in them. The novelty is to obtain quality aerospace parts that are also lightweight, reliable, and consistent. Today, it is highly common to use composites in energy, automotive, aerospace, sports, biomedical, and many other industries. Perhaps the most common and well-known complex lightweight structures using composite materials are the Boeing 787 "Dreamliner" (**Figure 1.1**) and the Airbus A350. It is estimated that composites constitute at least 50% of the weight of the Boeing 787, whereas they comprise only 12% of the weight of the Boeing 777. In addition, many of the composite parts in the 777 are non-structural, such as lavatory panels, whereas in the new 787, composite components are structural and load-bearing. Boeing, Airbus, and several other manufacturers used the opportunity to learn and gain experience in composites' design and manufacturing by integrating these non-structural parts in the airplanes with a clear vision to utilize their superior properties in later designs.

Figure 1.1 Breakdown of materials on the Boeing 787.

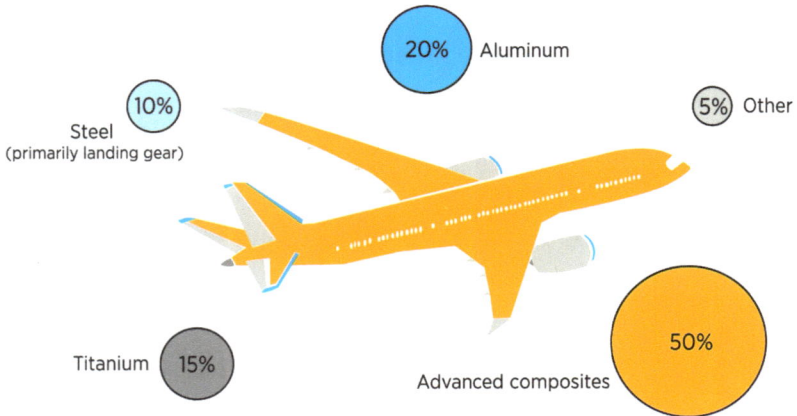

Aluminum 20%
Steel 10% (primarily landing gear)
Other 5%
Titanium 15%
Advanced composites 50%

Reprinted from Advances in Composites Manufacturing andProcess Design, F. Collombet,Y.-H. Grunevald,L. Crouzeix,B.Douchin,R. Zitoune,Y. Davila,A. Cerisier,R.Thévenin, Copyright 2015, with permission from Elsevier.

The use of composite materials can have a significant impact on the performance of the final product. While today the use of composites is often associated with additional complexity and cost in the design and manufacturing processes, the benefits during the operation of the product can outweigh these drawbacks. In the future, with improvements in our understanding of composite manufacturing technologies and refinements in design, more advancements can be made. Using the example of the Boeing 787 and 777 as a comparison, we will look at the impact of composites on the operational effectiveness and efficiency of the airplanes. As can be seen in **Table 1.1**, the 787 can transport one passenger for every 411 kg (compared to one passenger per 460 kg for the 777) and has an extended range of 11,910 km. This ability to transport people at less cost and with less impact on the environment not only has an economic impact but also positively contributes to the sustainability goals of most airlines.

Table 1.1 Key characteristics of Boeing's 777 and 787 Dreamliner (modernairliners.com).

Characteristic	777-300ER	787-10
Weight	167,800 kg	135,500 kg
Range	9700 km	11,910 km
Passenger capacity	365	330
Weight per one passenger	460 kg/passenger	411 kg/passenger

© SAE International.

Applications of Composites

Duration: 21 minutes

Description: This video, part of "Composites A-Z: 30 Days of Composites," covers the applications of composite materials. It highlights their strength, stiffness, and lightweight properties, making them ideal for aerospace, automotive, and other industries. It also discusses manufacturing methods, including hand layup and robotic processes, and provides case studies like the Boeing 787 and X-55 ACCA.

Scan the QR code below to watch the video.

Textiles and textile manufacturing are often seen as precursors to composite manufacturing. It is quite common for composite manufacturers to have their roots in the textile industry. Toray Industries, for example, which is today one of the major companies in the composite materials market, has its roots in textiles and fabrics. Recently, Toray acquired Tencate, another company whose roots are in the textile industry. Other major composite material manufacturers include Hexcel and Syensqo (formerly Solvay). Toray, Hexcel, and Syensqo dominate the composite materials industry and manufacture a wide range of woven fabrics, tape, slit tape (for AFP), adhesives, reinforcements, honeycomb structures, resins, and other composite constituents or products.

Did You Know?

While composites are one of the main reasons that the 787 Dreamliner is highlighted in this textbook, another important aspect would be the collaborative manufacturing nature of complex products. Today, most aircraft are built on partnerships extending beyond geographical boundaries using intricate product lifecycle management (PLM) systems to manage the thousands of parts involved.

vaalaa/Shutterstock.com.

1.2.
Terminology in Composites

Recall that the distinct components of a composite material are typically identified as:

1. *Reinforcement*: provides the strength component to the final material.
2. *Matrix*: provides the shape and adhesion of the structure, while also contributing support against shear forces, a function that fibers, in their straight form, are generally less effective at.

The reinforcement element, as stated earlier, can be continuous fibers or chopped fibers (**Figure 1.2**). Manufacturing continuous fibers is more demanding and requires additional control and physical mechanisms to ensure the appropriate manufacturing process compared to chopped fibers. In addition, the applications of aligned chopped fibers and long (several inches) discontinuous fibers are limited. To illustrate and provide a better example, let us consider three variations in the fused filament fabrication (FFF) manufacturing process. The first FFF process we consider is the most common variation, where the nozzle simply extrudes the thermoplastic material and consolidates on contact to create the shape, layer by layer. Typically one material (e.g., ABS, ULTEM) is used in this process with no distinct phases (homogeneous), and therefore, it is not a composite material. The second process variation includes injecting the original polymer material with chopped particles (e.g., short carbon fibers), and during the extrusion process, the nozzle delivers both materials, in their distinct phases, to create the structure. This can also be achieved with pre-prepared filaments that include the chopped fibers already within the polymer or by mixing a single-component filament at the extruder stage with chopped fibers. This variation type does not require further modification of the delivery system, but rather some minor process parameter modifications that do not necessitate major machine adjustment. Let us now consider the third, more complex, process variation, where the nozzle must simultaneously extrude a continuous filament (e.g., continuous graphite fiber) while providing the surrounding matrix material (e.g., ULTEM). This variation type requires significant machine adjustments, and one cannot simply use the traditional mechanisms of FFF 3D printing. Several additional steps and capacities are required to enable the process to achieve the desired product properties, such as cutting and nozzle design adjustment.

Figure 1.2 Continuous filament and chopped fibers (flake/particle).

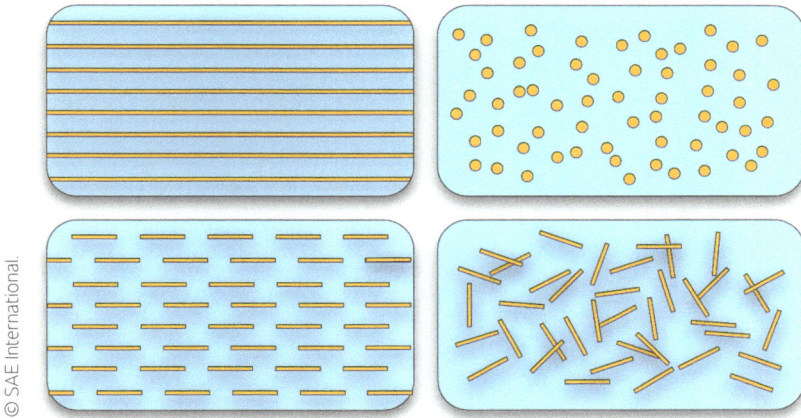

© SAE International.

Although the above examples employed polymers as the matrix element and carbon as the reinforcement, this combination is not an exclusive combination to form composites. Any engineering material can be used for either matrix or reinforcement. **Table 1.2** provides a list of different possible combinations. Although some of these combinations are highly uncommon for various reasons, including difficulty in manufacturing and/or cost, some of them dominate the composite industry. It must be noted that composites can always be created as long as each component is maintained in its distinct phase. However, the primary question is what variations are desirable from a functional, sustainability, or economic perspective.

Table 1.2 Possible reinforcement–matrix material combinations.

		Reinforcement		
		Metal	**Ceramic**	**Polymer**
Matrix	**Metal**	Metal–metal composites	Metal–ceramic composites	Metal–polymer composites
	Ceramic	Ceramic–metal composites	Ceramic–ceramic composites	Ceramic–polymer composites
	Polymer	Polymer–metal composites	Polymer–ceramic composites	Polymer–polymer composites

© SAE International.

1.2.1.
Material Forms

Many material forms used to manufacture composite parts are available to the design and manufacturing engineer. Below we list the most common ones and explain their predominant applications.

The first material form we discuss is *unidirectional tape* (**Figure 1.3**). When unidirectional tape is an input material for composite manufacturing, all the fibers within the tape layer have the same orientation. They are held together with the prepreg (matrix) material. An example of one of the most common unidirectional tapes is Hexcel's IM7-8552. The name of the material can provide insights into its different components. Typically, a carbon-epoxy material is identified first by the fiber and then by the matrix. For example, IM7 stands for "intermediate modulus" (IM) carbon fiber reinforcement material and 8552 is a thermoset resin system. It is essential to understand the resin system and its requirements (such as whether it is formulated to require an autoclave cure cycle or not). It is also important to be aware that unidirectional tapes are not all necessarily available in the form of thermoset prepreg composites. Recently, a new wave of thermoplastic resins (in contrast to the established thermoset resins), dry tape or low tack tape (that can be different material systems), have been introduced to industries. All these recent introductions still cannot compete with the established thermoset resin systems. They each have different niche applications and are often the result of innovative experiments geared toward specific goals: recyclability for thermoplastics, cost/shelf life for dry tape, machine performance for low tack, etc.

Figure 1.3 Unidirectional tape.

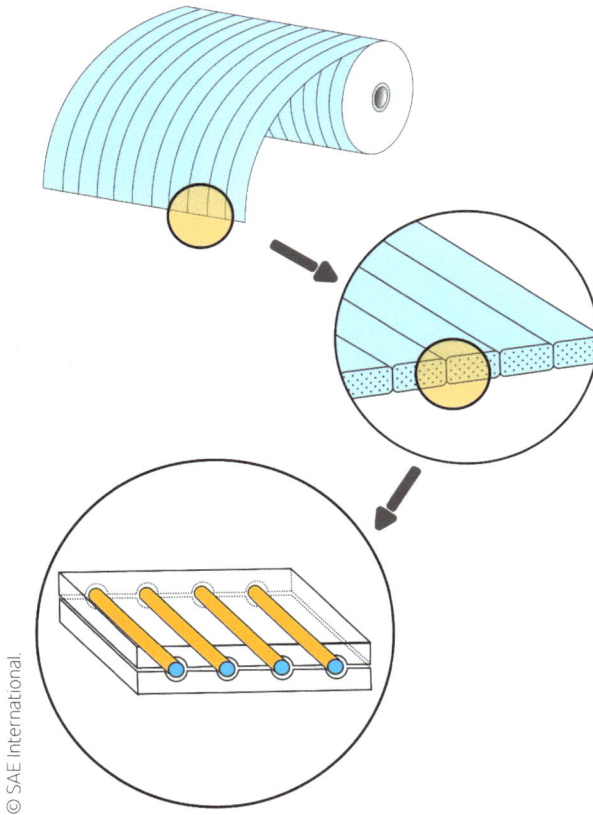

© SAE International.

Certain processes require the unidirectional tape to be slit into smaller dimensions, referred to as *slit tape* or *tows*, that can be used for a specific manufacturing process (**Figure 1.4**). Reasons include the creation of more intricate features and dimensions of machinery required. The fundamental difference between unidirectional tape and slit tape lies in the difference in the associated processes of automated tape laying (ATL, for unidirectional tape) and AFP (for slit tape). The dimensions of slit tape are typically ⅛, ¼, ½, and 1 in. in width. Unidirectional tapes generally have widths of 3 in. or more (typically up to 12 in.). Higher width rolls are typically used for hand layup only.

Figure 1.4 Slit tape of finite width depending on applications.

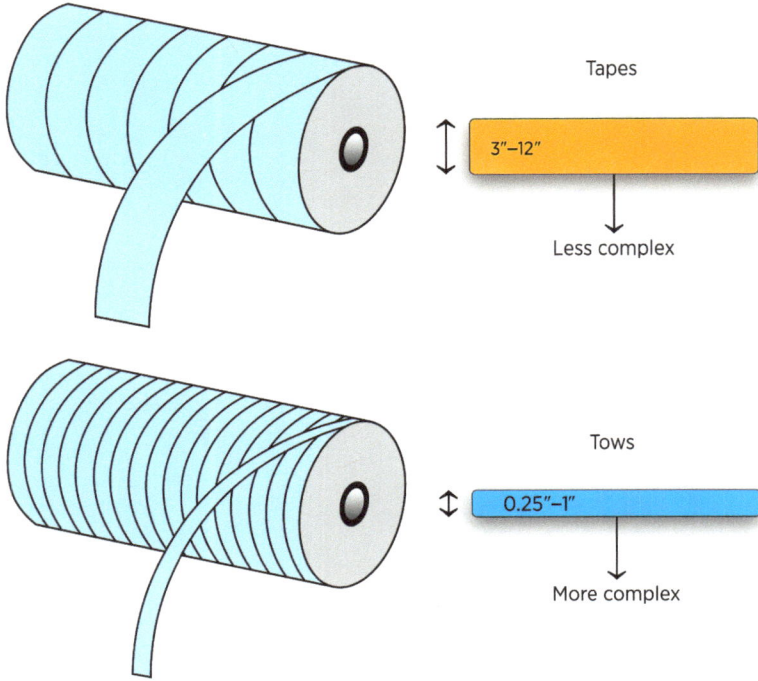

Tapes

3"–12"

Less complex

Tows

0.25"–1"

More complex

© SAE International.

Woven fabric is one of the most commonly used formats and is used for hand/manual layup techniques in composite manufacturing (**Figure 1.5**). As the terminology already indicates, woven fabrics are very similar to materials used in textile clothes. In woven fabrics, fibers oriented in two or more directions are typically integrated in the same ply, woven into each other. Woven fabrics do not necessarily require prepreg material and are often used as dry fibers. The resin system is applied in the second step to fill the space between the fibers of woven fabrics to create the intended shape as well as to help in shear.

Figure 1.5 Woven fabrics with two or more orientations in the same layer.

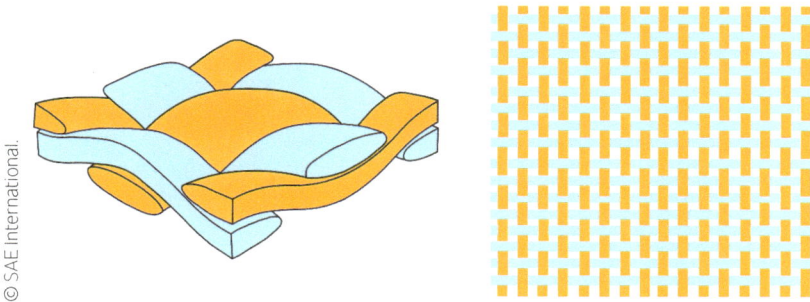

© SAE International.

To facilitate the use of composites, preforms that include several orientations of the in-plane fibers (reinforcement) are being created where the fibers are not woven together. Such material forms are called *non-crimp fabrics*. The layers of dry unidirectional fibers are stacked on top of one another and held together by a thin thread through the thickness. Theoretically, fibers of any orientation can be used in this arrangement. This grouping of fibers is carried out on a machine and serves as an intermediate material, decreasing the labor needed to place individual plies proportionally. This approach can also improve the quality by reducing human error such as fiber misalignment or missing layers.

Next, we will discuss *resin systems*. There are several variations when it comes to resin systems. Today, the aerospace industry predominately uses thermoset resin systems in primary aircraft structures. However, thermoplastics are likely to come into widespread use as soon as they are available at a competitive price and quality. They have a higher viscosity and therefore generally require higher heating and pressure cycles to achieve solidification compared to thermoset resin systems. For any resin system, one of the most important elements affecting the part quality that must be considered during the selection of a suitable resin system is the use of a proper cure cycle in terms of both temperature and pressure and the appropriate time constraints. Thermoplastics present the advantage of the out-of-autoclave (OOA) process, hence their desirability despite their higher heating requirements.

Did You Know?

The aerospace industry has been the primary user of composite materials since their inception. Such materials have been used in satellites, rockets, and aircraft applications. For example, the space shuttle used carbon–carbon panels on the nose and the wing leading edge to protect it from temperatures exceeding 2300°F encountered during reentry.

1.2.2.
Stacking Sequence

This section discusses what constitutes a "composite part," which is built from continuous fibers. In principle, we refer to a composite part as a collection of multiple plies stacked on top of each other in a layer-by-layer fashion. The creation of these plies is discussed later in this chapter. In this section, we present the stacking sequence and additional annotations that support this explanation.

First, let us introduce the essential concept of angle and orientation in the context of fiber placement. Any unidirectional tape has its "along the fibers" orientation. A single layer of unidirectional tape is referred to as a ply or lamina. These layers can be stacked to create a laminate where each ply may be oriented in a different direction. This varying orientation makes composites anisotropic materials, meaning the material properties are different in each direction. To define the ply orientations of the laminate, we set a reference orientation of 0° as the orientation along the longest side of the panel (or complex-shaped part) or, alternatively, as selected by design to achieve a specific property of the composite part. Subsequently, any other ply is labeled with respect to this original reference orientation. If the "along the fibers" orien-tation of a new ply deviates at a 45° angle from the reference orientation, this ply is classified as a 45° ply. Eventually, all individual plies are classified based on their angle with respect to the reference orientation. In the following example, the reference orientation is the bottom ply at 0° and all other plies are defined based on that orientation. An example stacking sequence is shown in **Figure 1.6**.

Figure 1.6 Example stacking sequence: [90/45/-45/0].

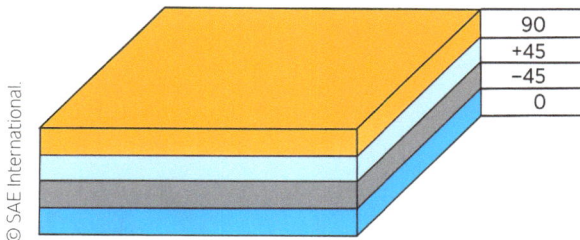

© SAE International.

At times, we insert padding, spacers, or local reinforcement, such as honeycomb layers, into complex structures between fibrous plies. Moreover, with recent advances enabled by modern manufacturing techniques, individual plies can have orientations that change within the plane of the ply, resulting in panels with variable in-plane stiffnesses. This in-plane change in orientation enables improved tailoring for specific locations in a structure where the loading and strength/weight requirements are well defined. Designing and manufacturing such parts is not trivial and only economically feasible at this point for specialty products, such as those used in aerospace or military applications. The sports industry utilizes variable stiffness in several applications, such as golf shafts with length-specific stiffness for optimized kick points, and skis/snowboards with tailored flexural responses near the edges.

We define the stacking sequence by listing the orientations from the first layer to the last sequentially. The stacking sequence for the composite example discussed below is first defined as [90, 45, −45, 0, 0, −45, 45, 90]. This nomenclature can be simplified by taking advantage of specific characteristics. If a composite laminate is symmetric, the subscript "s" can be used to simplify the annotation. Such laminates are desired due to the elimination of stretching-bending coupling in the ABD matrix. In the given example, since the composite is symmetric around its centerline, it can be annotated as $[90, 45, −45, 0]_s$. An additional simplification in the notation can be made when the same orientation occurs in the sequence for adjacent plies but with the opposite sign. In this case, the notation for two plies can be merged by introducing the "±" sign. The annotation is then $[90, ±45, 0]_s$ as shown in **Figure 1.7**.

Figure 1.7 Example stacking sequence: $[90, 45, −45, 0]_s$.

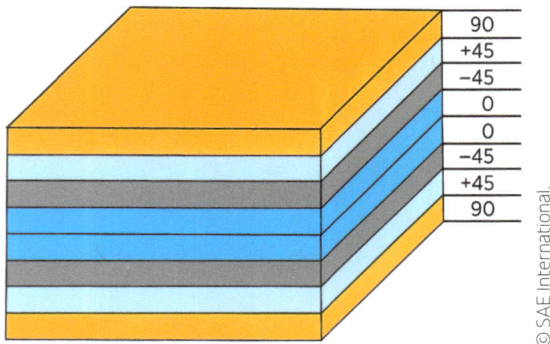

90
+45
−45
0
0
−45
+45
90

© SAE International.

Laminates are considered "balanced" when the laminate contains only pairs of plies at each orientation; for example, for each $+\theta$ ply, there must be a $-\theta$ ply. Specifically, balanced laminates represent laminates that have no extensional-shear interaction in the ABD matrix. The laminate discussed above includes a $-45°$ ply for each $+45°$ ply and is therefore considered balanced.

If the composite panel includes repetition, we can use parentheses and numerical subscripts to indicate this repetition. As such, [..., 90, 0, 90, 0, 90, 0, ...] becomes [..., $(90,0)_3$, ...]. If the composite includes an asymmetric middle layer, we can include a bar above that layer to indicate it is not repeated. As such, [90, 0, 45, 0, 90] becomes [90, 0, $\overline{45}]_s$. Additionally, though not required, sometimes a subscript "T" is added to the end of a stacking sequence to enforce the idea that the entire (total) laminate is defined so that the reader does not question whether an "s" subscript was merely forgotten.

A few examples of composite panels and their stacking sequence annotations are shown in **Figure 1.8**.

Figure 1.8 Example layup strategies top left: [45, -30, 30, -45], top right: [30, 90, -45, -30, 90, 45], middle left: [45_4], middle right: [45, 90, 0]_s, bottom left: [(0, 90)_3], bottom right: [-45, 60, -30, 45].

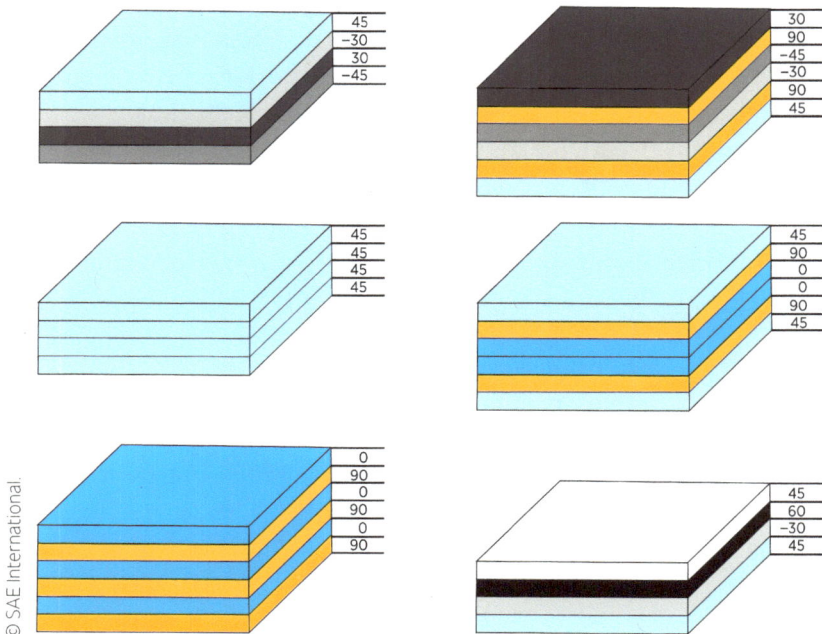

Material Forms

Duration: 23 minutes

Description: This video, part of "Composites A-Z: 30 Days of Composites," focuses on composite material forms. It covers unidirectional tapes, slit tapes, discontinuous fibers, woven and non-crimp fabrics, and sandwich panels. It highlights how different forms affect properties and performance, emphasizing that the best choice depends on application and manufacturing constraints.

Scan the QR code below to watch the video.

1.2.3.
Micro-Mechanics of Composite Structures

This section illustrates the different directional properties of composites. As previously stated, composites are anisotropic, and it is essential to understand the orientation of the material. In comparison, when designing a part using metals, one must consider only strength, weight, and volume. These features would be sufficient to analyze the design problem and attempt to solve it. For composites, several additional factors come into play that significantly affect the properties of the final structure. Therefore, design and manufacturing engineers need to consider these factors. For example, a flat panel with an unbalanced or unsymmetric stacking sequence will warp when cured and loaded. Careful attention to fiber orientations is necessary to ensure a suitable, composite-specific design is developed that is also manufacturable. One of the most common challenges within the composites' design process is that designers who have experience with metals must adjust their mentality to work with composites. It is no longer possible to propose a design without a solid understanding of the manufacturing techniques and their influence on the behavior of the final part.

1.2.3.1.
Theory

Composite manufacturing techniques are geared toward increasing the fiber volume while achieving adequate cohesion. To best illustrate this concept, we will annotate data relevant to the reinforcement phase by "r," to the matrix phase by "m," and to the overall composite by "c." Manufacturing techniques aim to eliminate air bubbles and voids trapped within a laminate. Unfortunately, such defects cannot always be avoided. The prevalence of these defects depends on the resin material and the resin curing process. Autoclaves, employing a cycle of pressure and heat, are most suitable to successfully reduce voids to a non-meaningful portion (<1% of overall volume). In general, we account for the volume of voids, using the subscript "v."

The composite total mass (m) and volume (V) are then computed as follows:

$$m_c = m_r + m_m$$

$$V_c = V_r + V_m + V_v$$

1.1

If an advanced manufacturing technique is properly used, such as employing an autoclave or an innovative OOA process, we can update the above equations by assuming the void volume is less than 1%, as follows:

$$m_c = m_r + m_m$$

$$V_c = V_r + V_m$$

1.2

1.2.3.2.
Volume Fractions

We now introduce the concept of volume fractions defined by the volumetric ratio of individual constituents with respect to the total volume. Designers aim to achieve the highest fiber volume structure possible while ensuring the matrix material is doing its job of providing appropriate cohesion to the fibers. Considering the above definition, we compute the volume fractions, annotated by f, as follows:

$$f_m = \frac{V_m}{V_c}$$

$$f_r = \frac{V_r}{V_c}$$

1.3

Considering the prior condition of a "no void" (=1% void space) composite, the volume fractions can be expressed as follows:

$$f_m + f_r = \frac{V_m}{V_c} + \frac{V_r}{V_c} = \frac{V_m + V_r}{V_c} = \frac{V_c}{V_c} = 1$$

1.4

1.2.3.3.
Density of a Composite Structure

The density of a composite part ρ_c can be calculated using the rules of mixtures. Below, we prove the direct application of the rules of mixtures and present a small application example.

$$\rho_c = \frac{m_c}{V_c} = \frac{m_r + m_m}{V_c} = \frac{\rho_r \times V_r + \rho_m \times V_m}{V_c} = \frac{\rho_r \times V_r}{V_c} + \frac{\rho_m \times V_m}{V_c} = \rho_r \times f_r + \rho_m \times f_m$$

1.5

A composite structure made with IM7-8552 has fibers constituting 40% of its volume. To compute the density of the structure, we first retrieve the individual densities of the IM7 reinforcement and the 8552 matrix system. IM7 by Hexcel is a continuous, high-performance, IM, PAN-based fiber provided in 12k filament count tows. Its density is 1.78 g/cm³. The resin system, 8552, is a high-performance tough epoxy matrix for use in primary aerospace structures. This resin system is developed to operate in environments up to 250°F. The density of 8552 is 1.3 g/cm³.

For this specific manufacturing process, using the described materials, we achieved a 40% fiber volume. We can now compute the composite density as follows:

$$\rho_c = \rho_r \times f_r + \rho_m \times f_m = 1.78 \times 0.4 + 1.3 \times 0.6 = 1.492 \, g/cm^3$$

1.6

The density of the composite structure in this case is 1.492 g/cm³.

1.2.3.4.
Modulus of Elasticity

Prior to discussing the computation of modulus of elasticity, it is imperative to highlight that, unlike density, the modulus of elasticity is orientation dependent. First, let us present the concept of *orthotropic* composite materials. Orthotropic composites are a subset of general anisotropic composites. The properties of orthotropic composites can be derived based on the individual properties along the orientation of the fiber and its perpendicular orientation. This is true when we have two mutually perpendicular planes of symmetry in the material properties.

In these scenarios, the rules of mixtures can be applied along the orientation of the continuous fiber only, and other derivations—that we will not detail—are needed for the other orientation. One can imagine the required computations to be similar to joining two springs end to end or in parallel. The continuous/chain mode represents the final modulus of elasticity along the orientation E_c of the fiber, and the parallel arrangement represents the modulus of elasticity along the perpendicular orientation E_c^p.

Along the orientation of the fibers, we use the following computation:

$$E_c = E_r \times f_r + E_m \times f_m \qquad\qquad \textbf{1.7}$$

Along the perpendicular orientation of the fibers, we use the following computation:

$$E_c^p = \frac{E_r \times E_m}{E_m \times f_r + E_r \times f_m} \qquad\qquad \textbf{1.8}$$

Application

One of the major concepts in composites is to try and maximize the fiber volume fraction, which means how much fiber we will eventually have in a composite versus the amount of resin. To compute the fiber volume fraction of a composite with fibers, we usually look at the representative volume element.

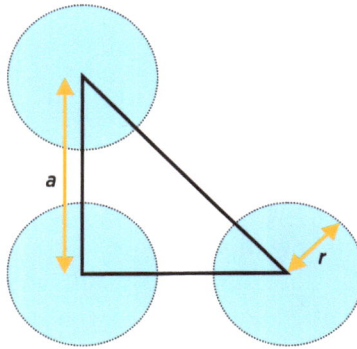

© SAE International.

What is the fiber volume fraction function of a and r?

$$V_c = \frac{1}{2} * a * a = \frac{a^2}{2}$$

$$V_f = \pi r^2 * \left(\frac{45}{360}\right) * 2 + \pi r^2 * \left(\frac{90}{360}\right) = \frac{\pi r^2}{2}$$

$$f_r = \frac{V_f}{V_c} = \frac{\pi r^2}{2} \bigg/ \frac{a^2}{2} = \frac{\pi r^2}{a^2}$$

What is the theoretical maximum value of the fiber volume fraction?

$f_{r,max}$ happens at $a = 2r$

$$f_{r,max} = \frac{\pi r^2}{(2r)^2} = \frac{\pi}{4}$$

1.3.
Composite Manufacturing

Composite manufacturing deals with different processes applied to composite materials to create parts and products. We can classify these processes as additive, deformative, or assembly. Some composite manufacturing experts classify composite manufacturing as a subset of additive manufacturing, especially when dealing with the laminated object concept. Others choose to classify some advanced concepts, such as AFP, as a microwelding process.

Metals are isotropic since they exhibit the same physical properties in all directions. Composites are anisotropic in that they exhibit different behaviors depending on the orientation. Specifically, layered composites are, by their very nature, anisotropic since they have different in-plane properties compared to their thickness direction. Additionally, since fibrous composites typically have fibers in different layers in different directions, each layer has different properties, such as stiffness and strength. These characteristics mean that the anisotropic nature of composite parts must be carefully considered in their design and manufacturing. Quasi-isotropic laminates, often dubbed "black aluminum," offer a compelling middle ground in composite design. They mimic isotropic material behavior by stacking plies in a balanced and symmetric manner, typically with orientations like 0,90, ±45. This eliminates directional bias, allowing engineers to apply isotropic analysis methods, simplifying design processes. They provide a practical way to achieve CFRP's weight advantages without extensive modifications to existing isotropic design frameworks.

The manufacturing of composite panels, due to the anisotropic properties of composite materials, requires close attention to orientation. Orientation is more critical for parts with continuous fibers compared to those with short fibers and particles. There are multiple composite manufacturing techniques that are dedicated to specific structures. The development of these processes was aimed at achieving specific requirements for specific designs. In the following, we will present principal composite manufacturing techniques that are widespread. There are several other techniques under development in advanced composites. These techniques will be added to subsequent editions of this book once they have reached a certain stage of maturity and relevance for industrial applications. In this section, we present the first seven subsections as shown in **Table 1.3**, whereas in the following section, we focus in greater detail on AFP.

Table 1.3 Overview of composite manufacturing techniques [1.1].

Molding method	Production rate	Capital equipment cost	Tooling cost	Labor cost	Finished part cost
Hand/spray layup	Low	Low	Low	High	Mid
Filament winding	Mid	Mid-high	Mid	Low	Low
Vacuum infusion processing	Low	Low	Low	High	Mid
Resin transfer molding	Mid	Mid	Mid	Mid	Mid
Compression molding	High	High	High	Low	Low
Pultrusion	High	High	High	Low	Low
Automated fiber/tape placement	Mid-high	High	High	Mid	High

Courtesy of American Composites Manufacturers Association.

1.3.1.
Hand Layup

Hand layup is the manual stacking of composite layers according to the design. It is the most common and widespread manufacturing technique for composites at present. Individual layers are cut from woven or unidirectional sheets and are positioned to create the laminate. The typical steps in hand layup are as follows: (1) individual layer preparation, (2) bagging material preparation, (3) projection of boundary, (4) positioning of materials, (5) application of resin system, (6) sealing, and (7) curing.

First, we prepare all the single layers. These layers are not necessarily of equal dimensions and can be of different sizes. It is very common to include pad-up reinforcements between layers and at certain specific regions, which results in a part of non-uniform thickness. Such built-up regions might be useful if holes are to be drilled or large cutouts are planned in late structural assembly. Note that the materials do not have to be similar. This step can include the preparation of honeycomb or other lightweight inserts as well. These would act as a core in a sandwich structure if placed in specific locations to reduce the weight of the overall part or product.

The next step is tool preparation. It typically involves preparing the bagging material and placement on the tool surface or placing mold release agents. For bagging, we include a first layer of the bagging material, such as a nylon bagging film, that is designed to be used in the curing cycle, which could be in an autoclave or oven, and could use an RTM process. It is important to understand the operating conditions of the bagging material to avoid burning or otherwise damaging it in the subsequent steps.

The third step is not required for all industries and applications. However, it is important to have support from a projection system that shows the exact location of each individual layer to increase precision and repeatability for selected critical applications. The projection of layer boundaries on the table or mold facilitates the reduction of human error and can support a more time-efficient production process. For some parts and products that require certification, the use of a projection system can be a factor supporting the fulfillment of certification requirements since it leads to more consistent parts.

The fourth step is the position of the fabric or ply cutout. If a projection system is used (see step 3), then the operator places the single layer prepared in step 1 within the boundary provided. If no projection system is available, the technician must place a single layer based on the available documentation, often in the form of drawings or layup plans (paper or digital).

In step 5, the resin system will be applied by the operator if the material being used is not a prepreg. Often, a hand roller is used to ensure equal repartition of the resin in most cases. Steps 4 and 5 are continuously repeated as needed to build the laminate layer by layer. Each layer must be placed precisely on the layer below, including all fabric, adhesives, core, etc. As this involves manual operations, steps 4 and 5 can lead to significant quality deviations depending on the operator's performance. This is one of the major critiques toward manual layup and a motivation to develop more automated, and thus controlled, composite manufacturing processes.

Finally, we seal the bagging process in step 6 and proceed with curing the composite in the final step (step 7) typically in an autoclave or oven. The usage of a peel ply is important to remove the part from the tool. We show an example of the hand layup process using fiberglass in **Figure 1.9**.

Figure 1.9 Example of the hand layup process with fiberglass.

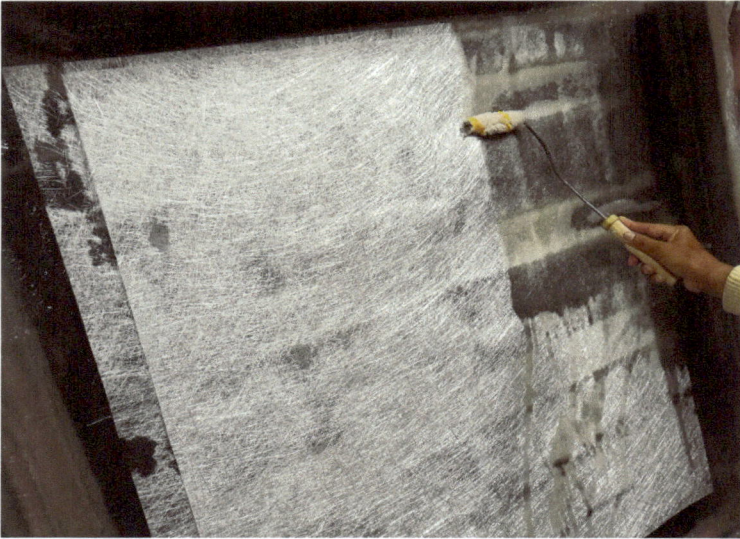

TASER/Shutterstock.com.

Scales

Duration: 30 minutes

Description: This video, part of "Composites A-Z: 30 Days of Composites," discusses micro and macro scales in composite analysis. It covers micromechanics, volume and mass fractions, stiffness, compliance matrices, and stress-strain coupling. The focus is on understanding how fiber and matrix properties influence composite behavior, with applications in predicting material performance.

Scan the QR code below to watch the video.

1.3.2.
Spray Layup

This manufacturing process is versatile and can be used for a wide variety of sizes and shapes. Also known as chop-spray layup, composite parts are created using a chopper gun. The latter cuts fibers into short lengths and mixes them with the desired resin prior to being sprayed onto the mold. A release agent is typically used whenever a mold is required to ensure the finished part will be easily removable. Spray layup is a cost-effective technique that enables high production rates, whereas surface finish is typically rough. One of the major disadvantages of spray layup is the harmful fumes and dust generated throughout the process. This requires additional safety measures and precautions (**Figure 1.10**).

Figure 1.10 Schematic of spray layup.

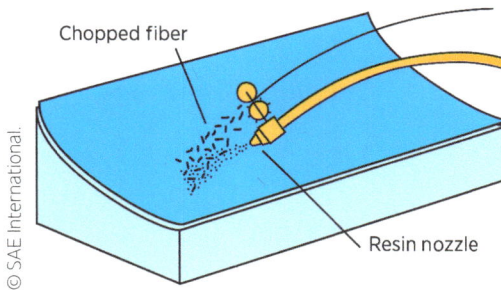

1.3.3.
Filament Winding (FW)

This manufacturing process is particularly suitable for pressure vessels and structures with circular cross-sections. It is rather simple: the structure is placed on a rotating mandrel that simultaneously pulls the fibers and places them. A simple manufacturing approach that ensures a uniform cross-sectional area in the composite parts is a good starting point. These continuous filaments (reinforcement material) are guided from a creel onto the structure. In between, they are impregnated with the resin (matrix material) by dipping them into a resin bath in situ (in process). The creel holds the fibers in their dry format, without any wetting, or with only limited toughening agents added. The separators ensure uniform spreading for fiber impregnation, and the final guiding ring merges the fibers prior to being placed on the structure as shown in **Figure 1.11**.

Figure 1.11 Schematic of the FW process.

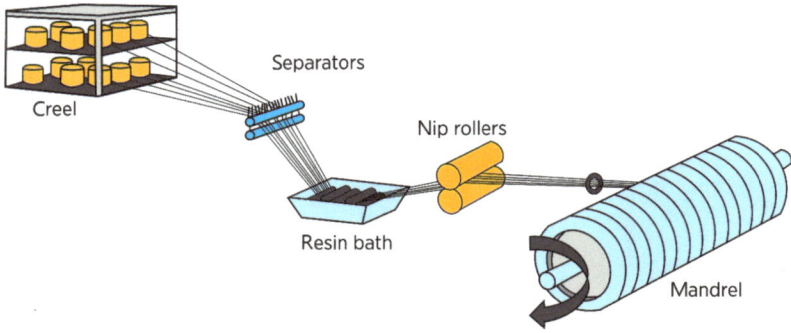

© SAE International.

Multiple proven layup strategies and patterns for FW are available. One of the most common ones is labeled hoop winding, where the filament is placed circumferentially at almost 90° to the axis of the part. Also, particular attention is needed at the tips of the structure (end closures). For structures that are designed to operate under specific pressure conditions, we can expect the end closures to be mechanically fastened to the filament wound structure.

Did You Know?

Pressure vessels are designed to contain liquids/gases at substantially high pressures. Composite-manufacturing-enabled composite overwrapped pressure vessels have far-reaching benefits as they enable space vehicles to travel further and save weight.

YAKISTUDIO/Shutterstock.com.

1.3.4.
Vacuum-Assisted RTM (VARTM)

The VARTM process is an OOA closed mold manufacturing technique. Due to the need for sufficient resin flow, a low-viscosity resin is typically used. Firstly, the layers of dry fiber fabrics are cut to the desired geometry and placed in the appropriate stacking sequence. Either a resin port or spiral tubing along with a vacuum port is then inserted into the desired location within the layup. A flow medium is then placed to assist with the flow of resin. Next, a peel ply layer is placed. This layer ensures clean removal of the bagging material after the infusion process. A vacuum bag is then placed to cover the previously placed layers. This vacuum bag is sealed to the surface with a sticky clay-like material, often referred to as "tacky tape." Vacuum is then applied to the system, and any leaks are found and fixed. Note that if leaks are not mitigated, proper resin flow and consolidation pressure will not be supplied. The location of resin and vacuum ports is also a critical choice. Improper selection of such locations can lead to air pockets in the laminate where no resin is transferred. Typically, simulations are performed beforehand to ensure adequate resin flow. After the resin infuses completely across the laminate and is allowed to cure, the bagging material can be removed, and the final part can be assessed. The VARTM process shown in **Figure 1.12** is deceptively simple but is difficult to master. The process is considered closed mold although the process has only one rigid mold half. The usage of the vacuum bag creates the "second" half of the mold.

Figure 1.12 Example of a VARTM setup.

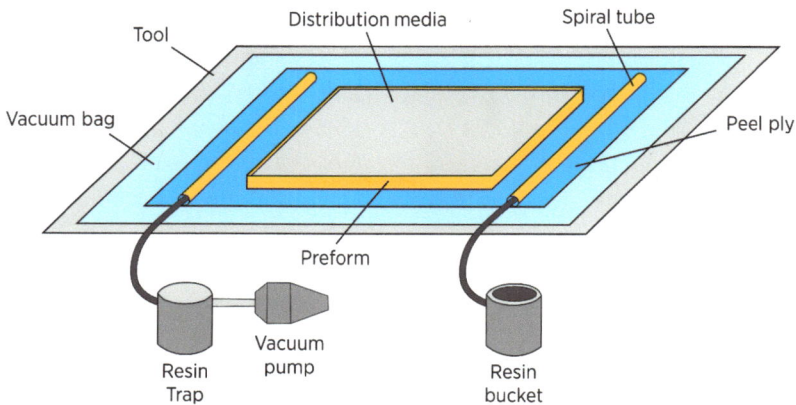

1.3.5.
RTM

The RTM process is an OOA process widely used in aerospace and other industries. The primary concept is that fiber preforms are shaped by an upper mold closing on a lower mold with a certain clearance. A mixing head with an intake of both resin and catalyst would pump into a resin injector that would introduce the matrix/resin to the mold. The process is also prone to the addition of cores, and complicated shapes, in successive stages, which is shown in **Figure 1.13**.

Figure 1.13 Example of the RTM process.

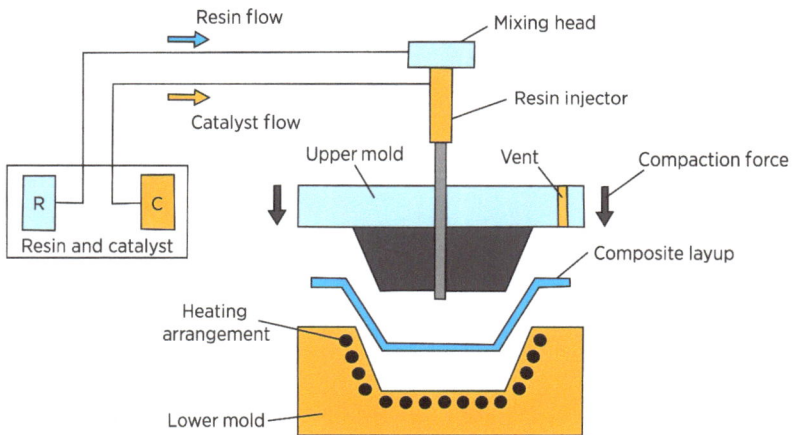

The RTM process is known for its ability to consistently produce top-quality components with excellent dimensional accuracy, and it has been proven to achieve this while maintaining strong mechanical properties. In contrast to VARTM, RTM does not use vacuum, typically beneficial for reducing voids. Compared to other composite technologies, RTM is known for its relatively low cost, which is one of its significant advantages. Applications of RTM range from biocomposite, aerospace, automotive, energy, and construction materials.

1.3.6.
Compression Molding

Compression molding is similar to die casting processes, in the sense that a composite preform is pressed along a mold and heated to initiate a curing process for a thermoset prepreg or to consolidate thermoplastic laminates (**Figure 1.14**). Following this step, we demold the part and retrieve it for further finishing processes such as drilling and trimming.

Figure 1.14 Schematic of the compression molding process.

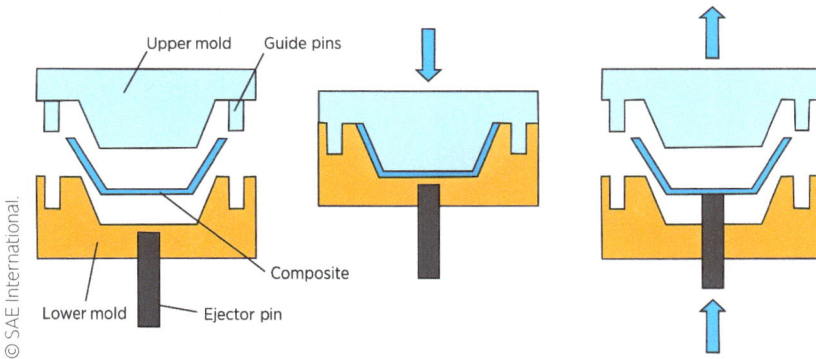

Upper mold Guide pins

Composite

Lower mold Ejector pin

© SAE International.

1.3.7.
Pultrusion

The term "pultrusion" is a hybrid between pulling and extrusion. The fibers are pulled in a manner similar to wire and bar drawing, i.e., from the exit point. Pultrusion attempts to join several steps into one manufacturing process. This inherently creates a manufacturing quality inspection issue, as it can be hard to know at which step a problem arose. The two primary steps that pultrusion combines are the impregnation of fibers as they are withdrawn from racks and routed through guides and the collimation of fibers into aligned bundles prior to entering the die which provides the profile/shape.

1.3.8.
ATL

First reported in the 1960s, ATL refers to the placement of a single tape onto a flat or slightly curved tool (**Figure 1.15**). The term "tool" is frequently used interchangeably with "mold" in the context of AFP/ATL, essentially referring to the same thing. It is the forerunner of AFP and shares similar steps: preparing the tape, programming the machine's path, prepping the tool surface, and then placing the tape. This results in a green state composite (for thermoset resins). Curing, demolding, and final finishing complete the process, as illustrated in **Figure 1.16**.

Figure 1.15 An ATL machine by Ingersoll Machine Tools creating a slightly curved part.

Courtesy of Ingersoll Machine Tools.

Figure 1.16 ATL process.

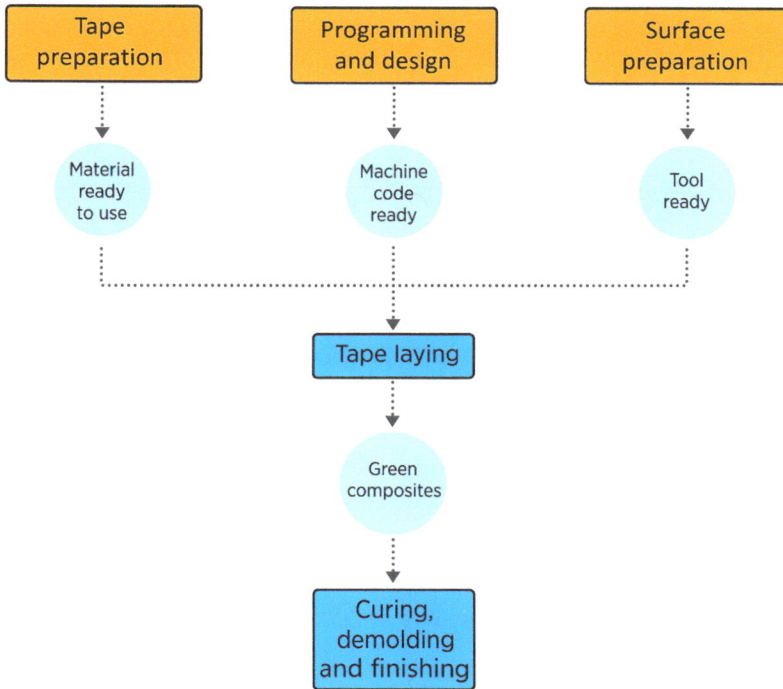

Tape preparation → Material ready to use

Programming and design → Machine code ready

Surface preparation → Tool ready

→ Tape laying

→ Green composites

→ Curing, demolding and finishing

© SAE International.

Did You Know?

As with any material, there are key properties that will be used to define and analyze composite materials. The most basic of these are elastic modulus (E), shear modulus (G), and Poisson's ratio (ν).

The elastic modulus, or modulus of elasticity, is a measurement of an object's resistance to being deformed elastically subject to stress. This value is determined by the slope of the stress–strain curve. A stiffer material will have a higher modulus and, therefore, more resistance to deformation. To better illustrate, diamond has a typical Young's modulus of 1220 GPa, while nylon has typically 2.7 GPa.

The shear modulus, or modulus of rigidity, is similar to the elastic modulus; however, it describes an object's tendency to shear when subject to opposing forces. This value is defined as the ratio of shear stress to the shear strain.

The Poisson's ratio, a measure of the Poisson effect, is a measure of the deformation of a material in directions perpendicular to the direction of loading. In other words, the expansion or contraction of an object in the width direction when loaded along its height. Most materials have values ranging between 0.0 to 0.5. The Poisson's ratios of rubber materials are closer to 0.5, while those of typical solid materials range from 0.2 to 0.3.

As we delve deeper into this book, you will discover that the seemingly straightforward properties we have discussed become much more intricate when dealing with composites. A key concept we will explore is directionality, where parameters like E (Young's modulus) will vary depending on the orientation, leading to distinct values like E_1 and E_2. Get ready to uncover the fascinating complexities of composites in our upcoming "Did You Know?" sections!

1.4.
AFP History

Previously, composite production of large structures was largely accomplished with ATL and FW, which became the precursors to AFP. The earliest documented account of using tows instead of ATL tapes was a patent by Goldsworth et al. in 1974 [1.2]. This invention utilized a splitting mechanism on an ATL head that slit 76.2-mm-wide tapes into 24 individual strands, now referred to as tows. The use of tows allowed for layup on increasingly complex parts that was not previously possible with wider tapes. Hercules Aerospace began the development of AFP machines in 1980, and they became commercially available later that decade, being implemented by aerospace companies such as The Boeing Company, Lockheed Martin Corporation, and Northrop Grumman [1.3]. The machines were, and still are, a combination of the differential payout capability of FW and the compaction and cut-restart capabilities of ATL. Developments of roller design, material guiding, and material heating from ATL were also directly applied to the AFP process. The AFP system had the capability to vary the layup speed, pressure, temperature, and tow tension. Bullock added to this capability by demonstrating an offline programming system that would benefit the machine's production time [1.4]. The offline system allowed the machine programming to be accomplished independently and then uploaded for execution.

Courtesy of Electroimpact.

A report described in 1993 [1.5], presented the implementation of a refrigerated creel system to prolong material life and allow for clean unspooling. Research in the 1990s was also focused on improving the productivity of the AFP process. This began with a system that could deliver up to 24 tows at once [1.6]. With this system, a layup rate of up to 30 m/min was reported, corresponding to a productivity of 1.9 kg/h, more than doubling the productivity associated with manual layup. Productivity continued to increase through reliability [1.7]. Reliability over complex geometries was improved by delivering tows along a curvilinear path, otherwise known as steering. An application of this development showed a 450% improvement in productivity, a reduced material wastage from 62 to 6%, and a cost reduction of 43% when compared with using a combination of FW and hand layup [1.8, 1.9]. These improvements in AFP also coincided with the development of thermoplastic composites for aerospace structural applications. The use of these materials allowed for in situ consolidation during layup, but higher placement temperatures and pressures are required [1.10]. Research on thermoplastic layups became a necessity due to the large size of composite structures exceeding the size of the autoclaves needed for curing [1.11].

Starting in the 2000s, a great deal of research was focused on improving process reliability and productivity. Boeing [1.12] and ElectroImpact (EI) [1.13] have performed studies on the amount of time delegated to manual inspection and rework of AFP layups. Boeing showed that layup inspection and rework comprised 63% of the cycle time to manufacturing a fuselage barrel, more than 2.5 times as long as the layup process. EI found that inspection and repair consumed 32% of the total time, while machine layup time consumed 27%. A 2006 patent was the first to describe an automated detection system [1.14]. This system would electronically access positional data to define a defect location, and then the machine would automatically return to that location. EI also made a major contribution to the productivity of AFP machines with the development of a high-speed system capable of 50.8 m/min with interchangeable heads and reduced tow-path length [1.15].

Research published in 2010 presented efficient simultaneous use of multiple machine cells as well as modular AFP heads [1.16]. The modular head offered advantages of 360° positioning, multiplicity of tow widths, short tow path, and offline maintenance. This was further enhanced in a report in 2013 with highly accurate robots demonstrating a three-sigma accuracy of ±0.08 mm [1.17].

The most recent industry-relevant AFP research topics consist of high-throughput AFP, minimal defect layups, and in situ thermoplastic layups. High-throughput AFP and minimal defect layups are focused on improving the overall quality and efficiency of AFP-manufactured structures. In situ thermoplastic layups are focused on combining layup and curing, preventing the need to perform a costly and size-limiting curing step. All the advancements presented in this section are summarized in **Figure 1.17**.

Figure 1.17 Timeline of AFP developments.

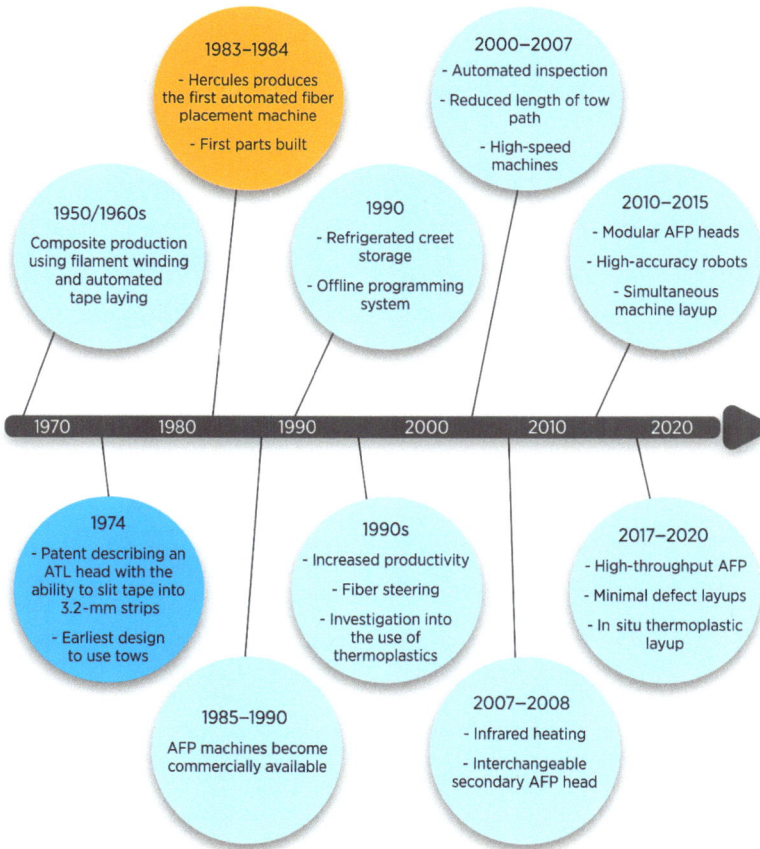

1950/1960s
Composite production using filament winding and automated tape laying

1983-1984
- Hercules produces the first automated fiber placement machine
- First parts built

1990
- Refrigerated creel storage
- Offline programming system

2000-2007
- Automated inspection
- Reduced length of tow path
- High-speed machines

2010-2015
- Modular AFP heads
- High-accuracy robots
- Simultaneous machine layup

1974
- Patent describing an ATL head with the ability to slit tape into 3.2-mm strips
- Earliest design to use tows

1985-1990
AFP machines become commercially available

1990s
- Increased productivity
- Fiber steering
- Investigation into the use of thermoplastics

2007-2008
- Infrared heating
- Interchangeable secondary AFP head

2017-2020
- High-throughput AFP
- Minimal defect layups
- In situ thermoplastic layup

References

1.1. American Composites Manufacturers Association (ACMA), *Basic Composites Study Guide* (Arlington: ACMA, 2017).

1.2. Goldsworth, W. and Hardesty, E., Geodesic path length compensator for composite-tape placement head. US Patent US3810805A, 1974.

1.3. Anderson, R.L., Grant, C.G., and NASA, "Advanced Fiber Placement of Composite Fuselage Structures," in *First NASA Advanced Composites Technology Conference, Part 2*, Seattle, 1991.

1.4. Bullock, D., "Automated Prepreg Tow Placement for Composite Structures," in *35th International SAMPE Symposium*, Anaheim, 1990.

1.5. Grant, C. and Benson, V., "Automated Fiber Placement - Evolution and Current Demonstrations," in *Third NASA Advanced Composites Technology Conference*, Long Beach, June 8–11, 1992.

1.6. Enders, M. and Hopkins, P., "Developments in the Fiber Placement Process," in *36th International SAMPE Symposium*, San Diego, 1991.

1.7. Evans, D., "Design Considerations for Fiber Placement," in *38th International SAMPE Symposium*, Anaheim, 1993.

1.8. Measom, R. and Sewell, K., "Fiber Placement Low-Cost Production for Complex Composite Structures," in *American Helicopter Society 52nd Forum*, Washington, DC, June 4–6, 1996.

1.9. Pasanen, M., Martin, J., Langone, R., and Mondo, J., "Advanced Composite Fiber Placement: Process to Application," 1997.

1.10. Cano, R.J., Belvin, H.L., Hulcher, A.B., Genoble, R.W. et al., "Studies on Automated Manufacturing of High Performance Composites," in *Structure Specialist Meeting*, Williamsburg, VA, 2001.

1.11. Gruber, M. and Lamontia, M., "Automated Fabrication Processes for Large Composite Aerospace Structures: A Trade Study," in *46th International SAMPE Symposium*, Long Beach, 2001.

1.12. Halbritter, A. and Harper, R., "Big Parts Demand Big Changes to the Fiber Placement Status Quo," in *SME Composites Manufacturing*, Mesa, AZ, 2012.

1.13. Rudberg, T., Neilson, J., Henschied, M., Cemenska, J. et al., "Improving AFP Cell Performance," *SAE Int. J. Aerosp.* 7, no. 2 (2014): 317-321, doi:https://doi.org/10.4271/2014-01-2272.

1.14. Engelbart, R., Chapman, M., Johnson, B., Soucy, K. et al., Systems and methods enabling automated return to and/or repair of defects with a material placement machine. US Patent US7039485B2, 2004.

1.15. Devlieg, R., Jeffries, K., and Vogeli, P., "High-Speed Fiber Placement on Large Complex Structures," SAE Technical Paper 2007-01-3843 (2007), doi:https://doi.org/10.4271/2007-01-3843.

1.16. Flynn, R., Nielson, J., and Rudberg, T., "Production Implementation of Multiple Machine, High Speed Fiber Placement for Large Structures," *SAE Int. J. Aerosp.* 3, no. 1 (2010): 216-223, doi:https://doi.org/10.4271/2010-01-1877.

1.17. Jeffries, K.A., "Enhanced Robotic Automated Fiber Placement with Accurate Robot Technology and Modular Fiber Placement Head," *SAE Int. J. Aerosp.* 6, no. 2 (2013): 774-779, doi:https://doi.org/10.4271/2013-01-2290.

A Coriolis C1 AFP machine configured for filament winding located at the National Composites Centre in the United Kingdom.

Chapter 02

AFP Process Parameters

AFP enables the complex manufacturing of advanced structures for several domains such as energy, sports, maritime, aerospace, and, more recently, general appliances. The benefit of having an extremely competitive strength-to-weight ratio is that it enables the entry of composites into new domains every day, from archery bows to multimedia equipment. While we will detail in future chapters the manufacturing imperfections, features, and defects that might arise from AFP, this chapter will present the fundamental process parameters that influence the manufacturing and placement of material. Highly repeatable and nearly defect-free fabrication of composite parts is critical to the success and widespread acceptance of composite materials. To achieve the desired surface coverage and ensure the manufactured parts closely match the original design, it is essential to carefully coordinate process parameters. These parameters should be tailored to the specific machine types and the unique combinations of matrix and fiber materials used in the part.

This chapter first introduces the different machine types and the nomenclature of parts within an AFP machine. Following this is the discussion of the four main elements that guide the manufacturing process: heat applied to support the cohesion/tackiness process or the consolidation, speed at which the machine operates, tow tension that ensures the tows are placed in their intended shape without roping, and compaction forces that support the adherence to the substrate. Whereas these four parameters are the principal process parameters that need to be investigated, we also present an understanding of environmental conditions needed for optimal manufacturing. It is important to note that this chapter references several AFP defects without detailing and explaining them. However, they will be discussed in the next chapter and can be used as a reference in parallel to reading this chapter.

2.1.
Machine Types and Nomenclature

As a cross between FW and ATL, AFP originally stood for advanced FW before being renamed to AFP, now an industry standard for the automation of manufacturing and fiber placement of complex contoured structures. The operation and overall design of an AFP system closely resemble that of an ATL machine. As an example, **Figure 2.1** illustrates a Mikrosam ATL machine; a simple modification to the head, switching from delivering tapes to tows, would effectively convert this equipment into an AFP machine. While the programming required for AFP is significantly more complex due to the intention of manufacturing intricate shapes, the fundamental machine structure remains largely the same. The goal of AFP is to enhance the precision and consistency of material placement by strategically depositing individual tows only in the necessary areas, utilizing advanced delivery mechanisms. This would then substantially reduce material scrap and the need to pay for expensive material that ends up not being used in the final part. AFP machines are typically multi-degree of freedom systems where the degrees of freedom assist in the ability to place material on varying tool surface geometry. The machines can be split principally into three main types: vertical gantry, horizontal gantry, and robotic systems. The following subsections will detail each.

Figure 2.1 Vertical gantry ATL machine by Mikrosam.

2.1.1.
Vertical Gantry

The most common type of AFP is the vertical gantry AFP and can be seen in many aerospace manufacturing facilities. The large gantry can support the manufacturing of sizeable structures. For example, an airplane wing or a long turbine blade for wind energy can move along the Cartesian frame to access and deliver as much material as the gantry can travel for. The size limitation of the part is only limited by travel rails built for the specific machine.

While the gantry supports the travel and delivery location, it is the end effector, or AFP head, mounted on it that facilitates the delivery of material. **Figure 2.2** shows the kinematics of a vertical gantry AFP system placing material onto a flat tool, and **Figure 2.3** shows a vertical gantry AFP machine by EI. Placement of material onto concave, convex, or doubly curved surface is also possible with such a system. With a rotator, to be described later, cylindrical and rotationally symmetric structures can be manufactured. The limitation of tool surface curvature is dependent on kinematics of the AFP machine and the clearance area around the AFP head. Some head designs are built to allow for better clearance, therefore enabling the placement of material on more severe curvatures. The manufacturing of complex geometries and how it connects to machine abilities will be provided throughout this book as AFP is intended to manufacture complex parts, not only planar or planar-like structures.

Figure 2.2 Vertical gantry AFP machine.

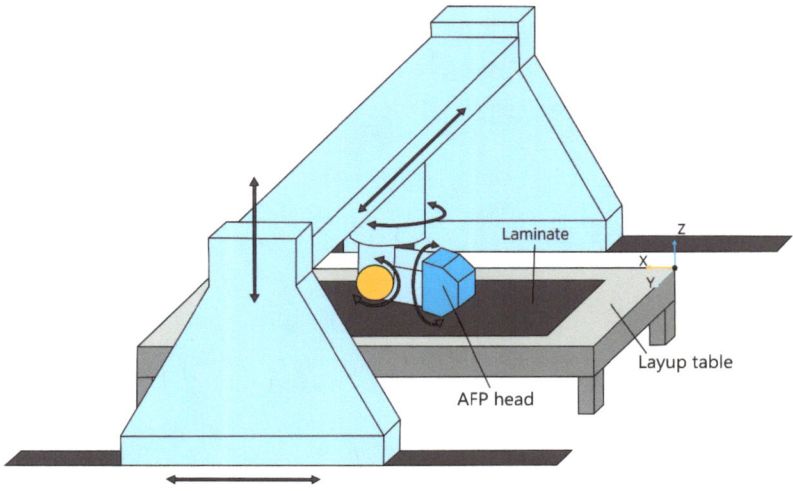

Laminate

Layup table

AFP head

Figure 2.3 Vertical gantry AFP machine by EI.

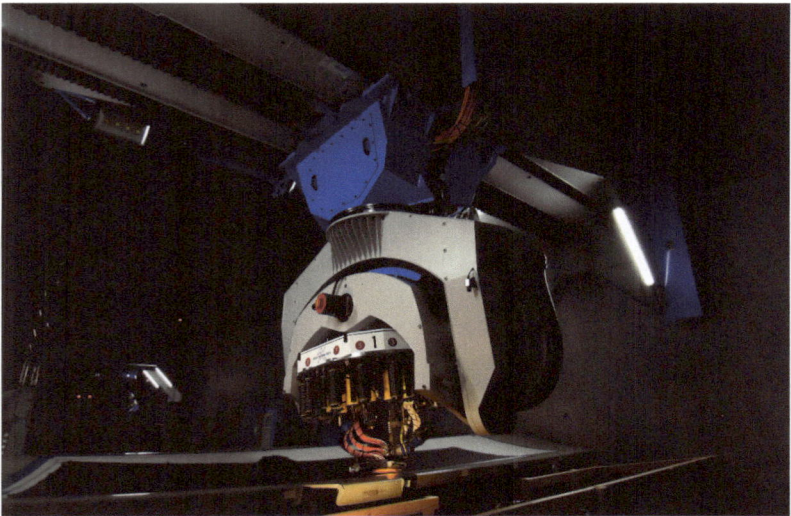

The head is usually mounted to a portion of the AFP system with three degrees of freedom, often rotational, that support attaining complex positions in a seamless manner. When creating a manufacturing plan, often referred to as optimization, it is crucial to account for the placement suitability of the toolpaths used to create the structure. As will be discussed in a dedicated chapter later, toolpath optimization is a critical step to ensure adequate manufacturing of the intended design. The differentiation between design and shape is important in the context of composites as the shape represents an external representation of the product, whereas the design includes the details of the fiber orientations along the layup path and is fundamental to assessing the part performance.

Did You Know?

Freebird7977/Shutterstock.com.

Composites are widely popular in the sports domain! The lighter weight advantage combined with the superior properties helped enhance performance and the player experience. Today, we can see golf club shafts, club heads, ball covers, and grips made out of composites in different frequencies. This supports an easier swing for golfers!

Using composites in the development of sporting goods provides a valuable platform for testing innovative ideas and refining manufacturing processes within a significantly shorter timeframe compared to industries like aerospace. For example, while the development cycle for a new aircraft design might take up to 30 years, creating a composite snowboard can progress from initial concept to on-snow testing within a couple of years. This accelerated development cycle allows for faster iteration and improvement of composite materials and manufacturing techniques. Once a new method or material has been successfully validated in the sporting goods domain, the knowledge and expertise gained can be transferred and applied to other industries, including aerospace, automotive, and construction. This cross-industry transfer of knowledge can lead to advancements in various sectors, driving innovation and improving product performance.

Vertical gantry AFP machines are particularly efficient for long parts that are not on a rotator. This does not negate the fact that rotators can exist in all types of machinery; rather, it is just what is more commonly available. Wing-like structures that span larger than 70 m are manufactured and inspected on such rotator-less types of machinery. Additionally, it is quite common for such types of machinery to have two creel delivery systems. Creels are refrigerated spaces where the spools reside prior to delivery. For thermoset materials, the creel is required to improve out-life time as the resin will begin to cure at room temperature. It also alleviates some tackiness of the material before deposition onto the surface, preventing issues such as foreign object debris (FOD), fuzz generation, tow stringers, and resin buildup inside the AFP head.

Two locations of creels exist that are pre- and post-motions. To better explain this concept, we can look to recent advances in AFP head assembly which enabled the creation of compact creel channeling where the spools are housed in a confined space by the delivery mechanism. The alternative is having the creel sitting on top of the gantry system to enable delivery through guiding rollers and mechanisms. With the increased distance, there is more likelihood of tow defects such as twists and folds due to the multiple redirects and extended tow paths. The increased amount of contact points also adds more places for resin and fuzz deposits that result in machine downtime due to the need for maintenance and cleaning. The deposits can also make their way through the delivery system and result in FOD deposits on the substrate. Such deposits require rework to remove the FOD and replace the affected tows.

Additionally, this setup enables crossover where the AFP head is swapped for an ATL head. The wider material used in ATL requires a different format of AFP head that cannot be integrated, with current technologies, with an AFP head designed for tows. The ability to switch heads is particularly useful for parts that are a hybrid between planar areas and certain complex zones. More planar areas can be placed with the ATL head, increasing the placement capacity of the system and resulting in lower process times. This is assuming that the switch of the head is not less efficient than the benefits of the increased placement capacity.

AFP/ATL

Duration: 29 minutes

Description: This video, part of "Composites A-Z: 30 Days of Composites," explores ATL/AFP, automated layup methods using prepreg for large structures. ATL uses wide tapes, AFP narrow tows, requiring skilled technicians for programming and maintenance. Key factors include hardware, process parameters, and defect management.

Scan the QR code below to watch the video.

2.1.2.
Horizontal Gantry

Horizontal gantry systems have the delivery head traveling on a XYZ gantry, but the delivery is horizontal in lieu of the vertical system previously discussed. **Figure 2.4** shows the AFP head with the capacity to deliver the head on a horizontal rotator to create full cylindrical-like structures or hoops. Boeing utilizes horizontal gantry AFP machines to manufacture the 787 barrel sections at their South Carolina facility. Horizontal gantry systems, such as the AFP machine at the McNAIR Center shown in **Figure 2.5**, are less common (recently) than vertical gantry, but they offer several advantages. One of those advantages is the ability to layup material on a tool that is adjacent to the structure without being in the same space envelope. This would allow for inline inspection on the other side of the layup, while not being in the layup domain/layup space of the fibers. The 787 barrel sections manufactured at Boeing in South Carolina use horizontal gantry AFP machines.

For illustration, we show the picture of the horizontal gantry system at the McNAIR Center in South Carolina where inspection and repair take place from the other side of the delivery system.

Figure 2.4 Horizontal rotator AFP machine.

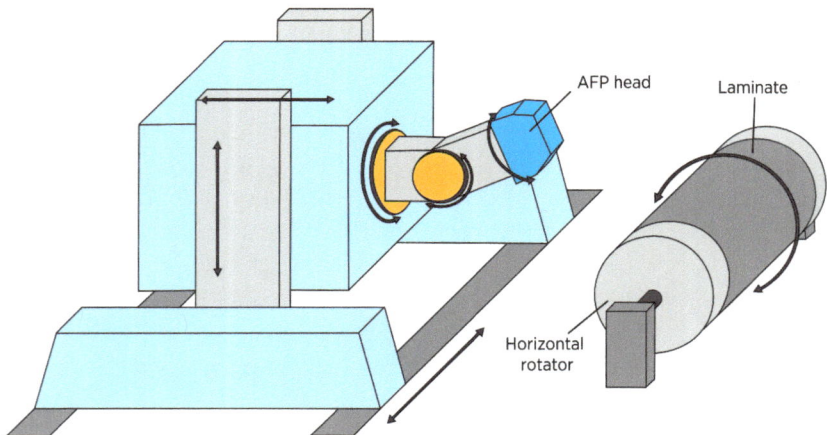

Figure 2.5 AFP machine at the McNAIR Center in South Carolina.

© SAE International.

2.1.3.
Robotic AFP

The accelerated adoption of AFP led to the need for more affordable systems that could be built faster with the same ability to manufacture complex shapes. This led to the development of robotic AFP platforms where the degrees of freedom for the fiber placement are provided by the robotic joints and incremented with the placement of the robotic arm on rails. **Figure 2.6** shows the AFP head mounted on a robotic platform to deliver the composite material and build the laminate on a layup table, and **Figure 2.7** shows a robotic AFP by EI.

Figure 2.6 Robotic AFP with layup table.

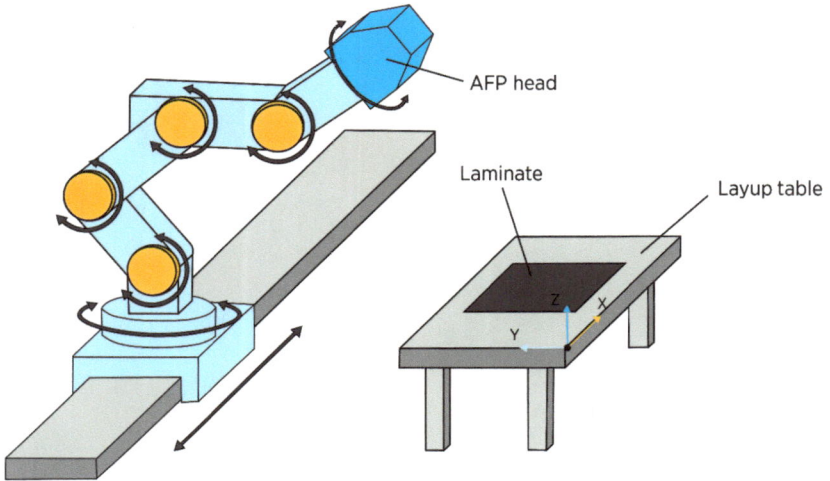

AFP head

Laminate

Layup table

© SAE International.

The same layup procedures, tables, and vertical/horizontal rotators are available with the robotic AFP platforms similar to gantry systems. The fundamental difference is that for almost all of these platforms, the material is located in the AFP head. These platforms are known for their versatility. Integrated Structural Assembly of Advanced Composites (ISAAC) is the AFP robotic platform at National Aeronautics and Space Administration (NASA) Langley. This robotic platform can swap the head for two diverse kinds of stitching heads, a projection head, and others. This makes robotic AFP ideal in research setup and industrial applications. ISAAC is an EI AFP machine with the capacity to simultaneously layup 16 tows to manufacture composite parts. **Figure 2.8** shows the robotic arm mounted on a robotic rail, while the AFP head is placing material on the tool. The figure shows the robotic rail, robotic arm, AFP head, and the tool used in the manufacturing of a blade-like structure [2.1]. Up to 16 6.35-mm (0.25-in.)-wide prepreg tows can be laid simultaneously by the AFP head. Following tape backing removal, prepreg tapes are redirected from spools located on the head into a feeding mechanism. The tows are fed onto the surface and cut to the appropriate length while

Figure 2.7 Robotic AFP by EI.

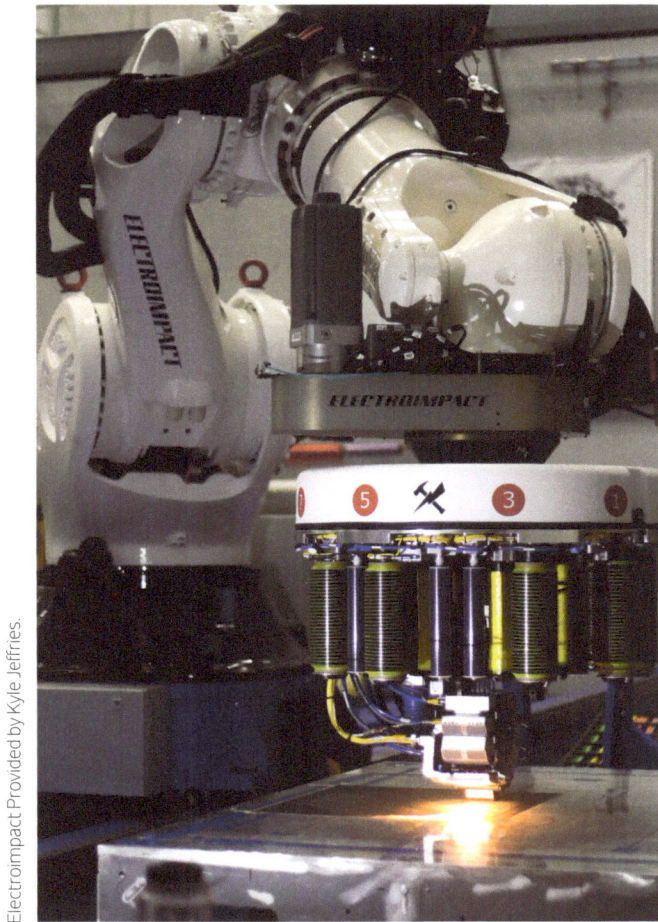

Electroimpact Provided by Kyle Jeffries.

being compacted by a 101.6-mm (4-in.)-wide polytetrafluoroethylene-coated roller. Several tows placed at once comprise a course, which can be placed both unidirectionally and bidirectionally. Multiple courses in one layer make up a ply, and several plies constitute the fully formed laminate. All aspects of the robotic system, as well as a 3.7-m by 1.8-m coupon table and additional end effectors, are located within the work cell. The entire system is located in an ISO 7 clean room to maintain control over environmental conditions and airborne particulates.

Figure 2.8 ISAAC at NASA Langley.

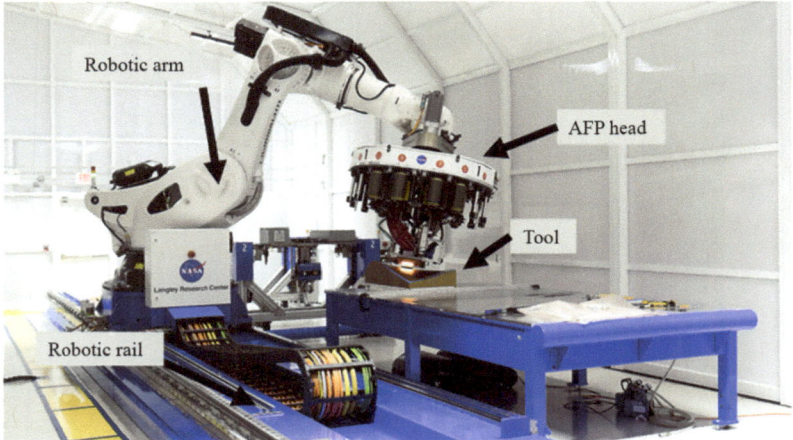

R. Harik, J. Halbritter, D. Jegley, R. Grenoble and B. Mason, "Automated Fiber Placement of Composite Wind Tunnel Blades: Process Planning and Manufacturing," in SAMPE Conference & Exhibition, Charlotte, North Carolina, US, 20 – 23 May 2019. Reprinted by permission from the Society for the Advancement of Material and Process Engineering (SAMPE).

Figures 2.9 and **2.10** show the placement and creation of a laminate on vertical and horizontal rotators, respectively. Horizontal rotators are more common as they do not limit the design space and can have the tool extend as much as the rail extends. **Figure 2.11** shows a horizontal rotator by Mikrosam.

Figure 2.9 Vertical rotator robotic platform.

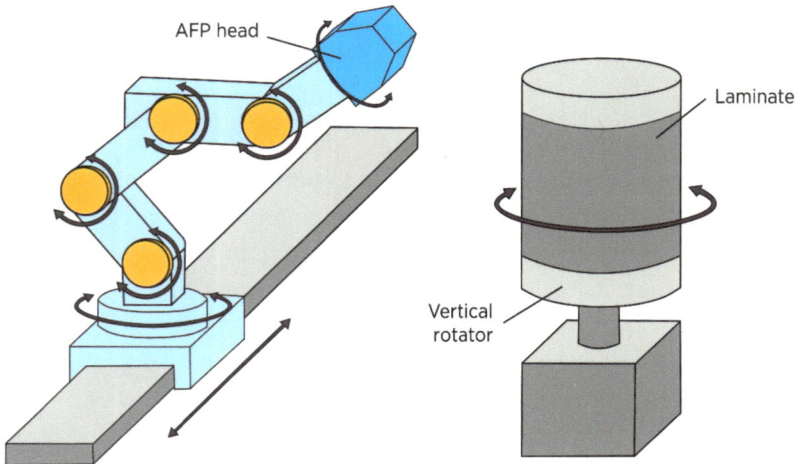

© SAE International.

Figure 2.10 Horizontal rotator robotic platform.

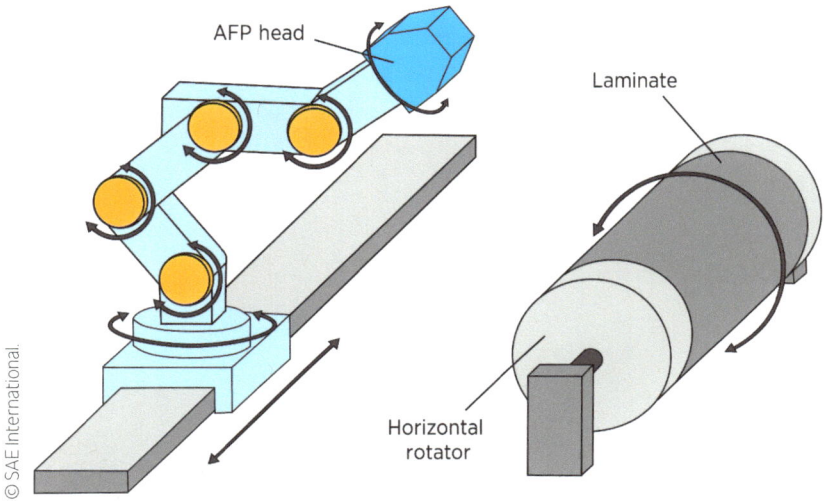

AFP head

Laminate

Horizontal
rotator

Figure 2.11 Mikrosam horizontal rotator.

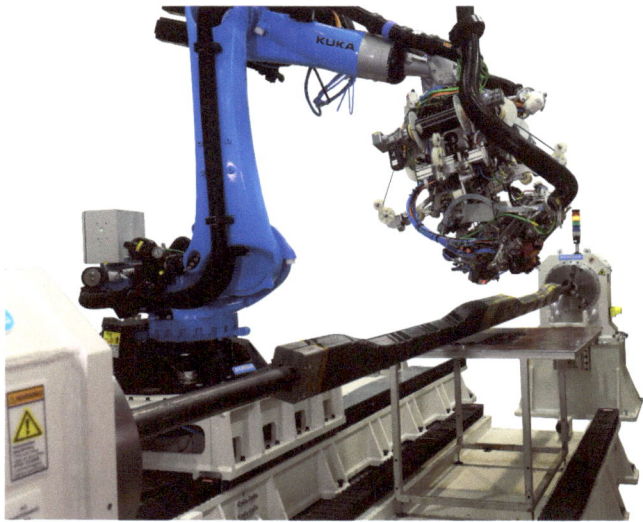

Recent advancements in AFP technology have introduced innovative infrastructure configurations. In some cases, the robotic arm holds the tool, while the AFP head remains stationary or rotates, as illustrated in **Figures 2.12** and **2.13**. Alternatively, dual robotic systems are employed for toolless manufacturing, as shown in **Figure 2.14**. These new configurations offer alternative flexibility in the AFP process.

Figure 2.12 Collaborative robotic AFP system.

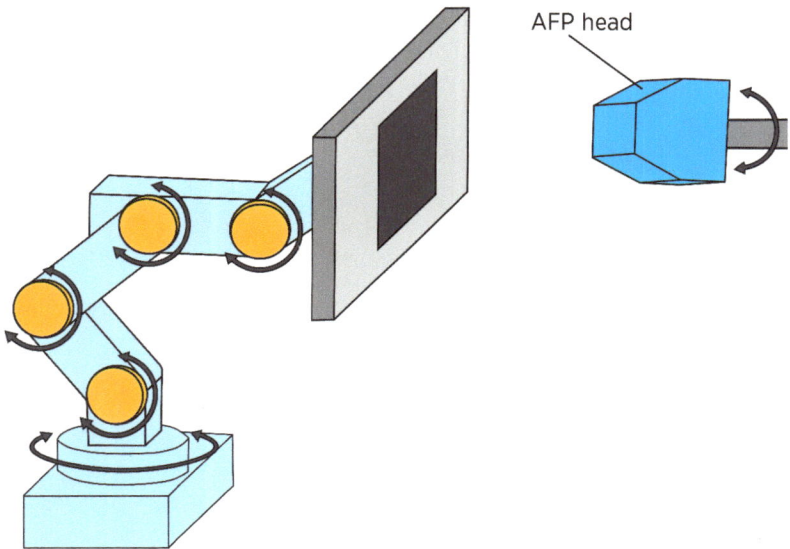

AFP head

© SAE International.

Figure 2.13 Ingersoll machine tools inverted robotic AFP.

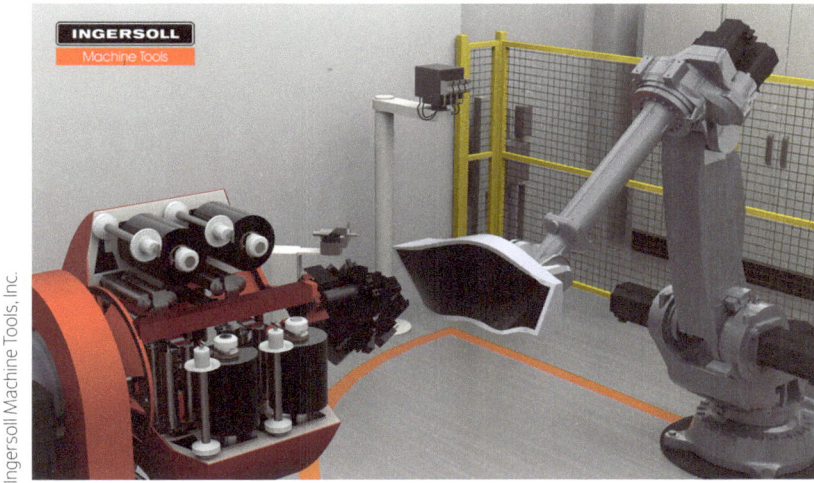

Ingersoll Machine Tools, Inc.

Figure 2.14 Mikrosam dual toolless robotic AFP.

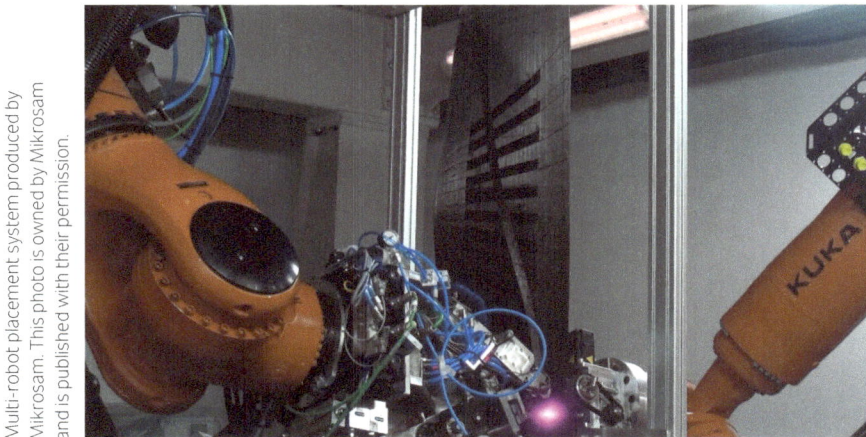

Multi-robot placement system produced by Mikrosam. This photo is owned by Mikrosam and is published with their permission.

2.2.
AFP Head

More recent AFP machines that have been developed utilize a modular AFP head, which has demonstrated high rate and high quality in commercial aircraft production [2.2]. The development of a modular head came from the desire to use multiple tow widths and enable offline maintenance [2.3]. This head offers advantages such as multiplicity of tow widths, very short tow path from the spool to the nip point, and rapid head change [2.4].

Figure 2.15 Schematic of an AFP head.

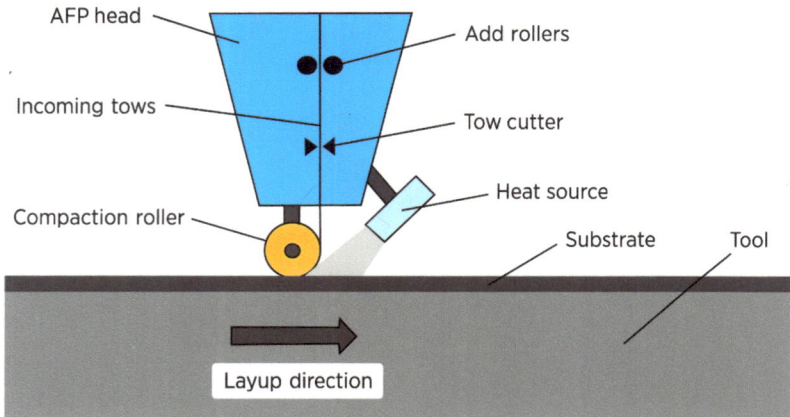

The AFP head (see **Figure 2.15**) contains all the elements needed to deliver the fiber, heat the underlying layer to provide for tackiness, and compress the newly delivered layer on the existing one. The heating/compaction mechanism has the in situ consolidation effect for thermoplastic resin material in lieu of the usual tackiness-only anticipation for thermoset resin matrix materials.

The AFP head typically has the following parts:

- Add rollers: They function as a mechanism to start the feeding. One of the main benefits of using tows is the ability to customize the layup and introduce tows only where they are needed. The added rollers act as a mechanism that compacts the rollers and drives the tows through the chutes and other mechanisms onto the tool.

- Incoming tows: These are the fiber tows that are delivered from the creel or from the positioned spools onto the different guides within the head leading to the tool.

- Tow cutter: One of the major advantages of AFP is the ability to place material only in places where it is needed. This provides the flexibility to add, cut, and restart at distinct locations to create patterns that are needed for the design at hand. One thing to understand is that this contributes to the creation of a minimum cutting length that we should understand in our design. This is what results in dog ears and other patterns that will be discussed in the following chapters.

- Heat source: The application of heat is a key factor in ensuring proper adhesion between the substrate and incoming tows. The apparatus that applies such heat is referred to as a "heater" or heat source. There are many hardware setups with each being beneficial for a given use case. These will be detailed in a later section.

- Compaction roller: Compaction pressure ensures appropriate placement of the incoming material and is supplied by a compaction roller. There are many variants of rollers that will be described in a later section.

Figure 2.15 also presents two other key terms. The first is the tool. In AFP terminology, the tool is the mold that acts as the recipient of the material. Tools are typically designed to withstand the compaction forces and the heat cycle from subsequent autoclave cycles (in the case of thermoset matrix resins). The other term is substrate. It is the buildup of material that is being delivered. Similar to work in progress (WIP), in traditional manufacturing, it is the material that has already been laid down and is waiting to receive a new layer of material.

Following this initial introduction to the complicated process parameters that need to be taken into consideration for composite layups. The next steps involve detailing the main process parameters and understanding their influence and criticality in obtaining the designed part. The endeavor of this section is to either (1) manufacture what was designed to ensure fidelity or (2) understand process parameters and how they influence the design, so that design changes or accounts for manufacturing in their process. This is also known as design for manufacturing (DFM).

Application

Traditional materials such as metals are isotropic materials, meaning they behave the same when loaded in all directions. This fact is not true for composite materials which, at best, can be approximated as orthotropic materials. Due to this attribute, it is necessary to compute stiffnesses and loads in specific directions. The following equations provide the reduced stiffness matrix for a given ply in a laminate.

$$\mathbf{Q}_{ij} = \begin{bmatrix} Q_{11} & Q_{12} & 0 \\ Q_{12} & Q_{22} & 0 \\ 0 & 0 & Q_{66} \end{bmatrix}$$

where

$$Q_{11} = \frac{E_{11}^2}{\left(E_{11} - v_{12}E_{22}\right)}$$

$$Q_{12} = \frac{v_{12}E_{11}E_{22}}{E_{11} - v_{12}^2E_{22}}$$

$$Q_{22} = \frac{E_{11}E_{22}}{E_{11} - v_{12}^2E_{22}}$$

$$Q_{66} = G_{12}$$

However, a laminate is made up of plies at various ply angles. Therefore, each stiffness matrix must be transformed into the global coordinate system at which the ply angles are defined.

This was found by computing the transformed reduced stiffness matrix.

$$\mathbf{\overline{Q}_{ij}} = \begin{bmatrix} \overline{Q_{11}} & \overline{Q_{12}} & \overline{Q_{16}} \\ \overline{Q_{12}} & \overline{Q_{22}} & \overline{Q_{26}} \\ \overline{Q_{16}} & \overline{Q_{26}} & \overline{Q_{66}} \end{bmatrix}$$

where

$$\overline{Q_{11}} = Q_{11}\cos^4(\theta) + 2(Q_{12} + 2Q_{66})\cos^2(\theta)\sin^2(\theta) + Q_{22}\sin^4(\theta)$$

$$\overline{Q_{12}} = Q_{12}(\cos^4(\theta) + \sin^4(\theta)) + (Q_{11} + Q_{22} - 4Q_{66})\cos^2(\theta)\sin^2(\theta)$$

$$\overline{Q_{16}} = (Q_{11} - Q_{12} - 2Q_{66})\cos^3(\theta)\sin(\theta) - (Q_{22} - Q_{12} - 2Q_{66})\cos(\theta)\sin^3(\theta)$$

$$\overline{Q_{22}} = Q_{11}\sin^4(\theta) + 2(Q_{12} + 2Q_{66})\cos^2(\theta)\sin^2(\theta) + Q_{22}\cos^4(\theta)$$

$$\overline{Q_{26}} = (Q_{11} - Q_{12} - 2Q_{66})\cos(\theta)\sin^3(\theta) - (Q_{22} - Q_{12} - 2Q_{66})\cos^3(\theta)\sin(\theta)$$

$$\overline{Q_{66}} = (Q_{11} + Q_{22} - 2Q_{12} - 2Q_{66})\cos^2(\theta)\sin^2(\theta) + Q_{66}(\cos^4(\theta) + \sin^4(\theta))$$

Using these equations, compute the reduced stiffness matrix and transformed reduced stiffness matrix for a 0-, 45-, −45-, and 90° ply with $E_1 = 20 * 10^6$ psi, $E_2 = 1.5 * 10^6$ psi, $G_{12} = 1 * 10^6$ psi, and $\nu_{12} = 0.34$.

Solution:

$$\mathbf{Q_{ij}} = \begin{bmatrix} 22.61 * 10^6 & 0.58 * 10^6 & 0 \\ 0.58 * 10^6 & 1.70 * 10^6 & 0 \\ 0 & 0 & 1 * 10^6 \end{bmatrix} \text{psi}$$

$$\mathbf{\overline{Q}^0} = \begin{bmatrix} 22.61 * 10^6 & 0.58 * 10^6 & 0 \\ 0.58 * 10^6 & 1.70 * 10^6 & 0 \\ 0 & 0 & 1 * 10^6 \end{bmatrix} \text{psi}$$

$$\bar{\mathbf{Q}}^{45} = \begin{bmatrix} 7.37*10^6 & 5.37*10^6 & 5.23*10^6 \\ 5.37*10^6 & 7.37*10^6 & 5.23*10^6 \\ 5.23*10^6 & 5.23*10^6 & 5.80*10^6 \end{bmatrix} \text{psi}$$

$$\bar{\mathbf{Q}}^{-45} = \begin{bmatrix} 7.37*10^6 & 5.37*10^6 & -5.23*10^6 \\ 5.37*10^6 & 7.37*10^6 & -5.23*10^6 \\ -5.23*10^6 & -5.23*10^6 & 5.80*10^6 \end{bmatrix} \text{psi}$$

$$\bar{\mathbf{Q}}^{0} = \begin{bmatrix} 1.70*10^6 & 0.58*10^6 & 0 \\ 0.58*10^6 & 22.61*10^6 & 0 \\ 0 & 0 & 1*10^6 \end{bmatrix} \text{psi}$$

Note the differences in the various transformed stiffness matrices. This gives you an understanding of how the plies behave when loaded outside of the fiber direction.

Thermoforming

Duration: 21 minutes

Description: This video, part of "Composites A-Z: 30 Days of Composites," explores thermoforming. A fast, closed molding process using heat, it produces strong, complex parts with prepreg sheets or chopped fibers (SMC/BMC). High upfront costs yield low part costs at volume, with cycle times under 1-9 minutes.

Scan the QR code below to watch the video.

2.3.
Heat

The application of heat is a key factor in ensuring proper adhesion between the substrate and incoming tows. The heater is a device that is mounted to the AFP head that supplies heat during deposition to ensure adhesion of the incoming tows to the substrate. The devices that will be covered consist of hot gas torches (HGTs), infrared (IR) heaters, lasers, and pulsed light heaters. These heating systems are summarized in **Table 2.1**.

Table 2.1 Summary of the characteristics of each heating system.

Heating system	Material	Controllability	Temperature
HGT	Thermoset/thermoplastic	Low	Medium
IR	Thermoset	Medium	Low
Laser	Thermoplastic	High	High
Pulsed light	Thermoset/thermoplastic	High	Medium/high

© SAE International.

Understanding the heat delivered to the substrate is crucial for an appropriate process development that is product specific. It is imperative to understand that the heating source acts in concert with other process parameters and that, for every product, a design of experiment with different levels needs to be conducted where the factors are heat, speed, compaction, tension, and environmental elements. The heat source selection also depends on the matrix resin system. Thermosets do not have a melting point; instead, they are characterized by a sub-ambient glass transition temperature (Tg) before curing (usually near 0°C) and a dry or wet Tg after curing (~120–200°C). Due to the low Tg prior to curing, thermosets typically require lower temperatures for placement, allowing the use of heating systems with lower applied heat. In many cases, IR heating can be sufficient for thermosets. However, thermoplastics require significantly higher heating due to their higher glass transition temperatures. In these cases, IR heating might not be sufficient, necessitating alternative heating methods capable of reaching the required temperatures for proper processing.

HGTs have been used for more than two decades and were used as heating mechanisms in ATL machines and in initial AFP machines [2.5].

This mechanism uses a hot gas, usually nitrogen, with the applied temperature being controlled by the gas flow rate. Using an HGT is comparatively inexpensive, but the temperature is difficult to control [2.6].

Figure 2.16 shows the placement of an HGT and the heating of the incoming tow. Several trials have been attempted with HGTs to study how many nozzles are needed to perform the function of heating, the incoming tows, and/or the substrate based on the best-case scenario. Additionally, early trials have attempted to have multiple heating systems acting simultaneously to achieve the desired temperature. This was not the case for thermosets but rather in the case of thermoplastic, where HGTs were used to heat up the incoming tows and IR was used to heat the substrate. These early equipment trials were not successful in achieving a production level; however, the research was a precursor to successful thermoplastic placement.

Figure 2.16 Schematic of an HGT heating system.

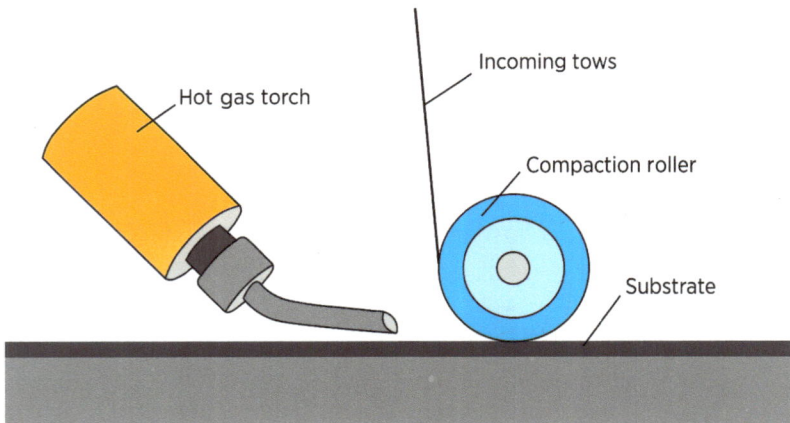

IR systems are commonly used in thermoset applications. Heat transfer from the IR heater to the substrate is done through radiation. It is also key to note that IR heating, as represented here, only applies heat to the substrate. We will see that other heating systems apply heat to both the substrate and the incoming material. When radiant energy hits an object, some of that energy is absorbed, some is reflected away, and some is transmitted through. The absorbed energy results in heating of the substrate, while the reflected energy is essentially wasted. For this reason, a reflector is often incorporated to ensure that most of the emitted energy is sent in a useful direction. The reflector sends the energy reflected toward the heating system onto the substrate again to be absorbed, increasing the heating efficiency. However, this heating system has a main disadvantage of inefficient heat transfer and non-uniform heating due to the wide dispersion of the heat [2.7].

Figure 2.17 shows one of the most common configurations for IR heaters where we have four bulbs that are heating the underlying layer in the substrate. These bulbs are typically operated individually, and we can have them operate at a percentage of their capacity. This usually is synchronized with the speed at which the layup is taking place, and also at times, we can emit a full intense heating cycle at a small-time interval. Bulbs closer to nip line, or line of contact between the incoming tows and substrate, are often supplied with high power leading to more heat. The variation in bulb irradiance prevents overheating of the substrate. Recent research is trying to move from an open-loop system, where the heating is not validated/controlled, to a closed-loop system. This is accomplished by using sensors that measure the heat on the surface and use this information to control the power being supplied to each bulb [2.8].

Figure 2.17 IR heating system schematic.

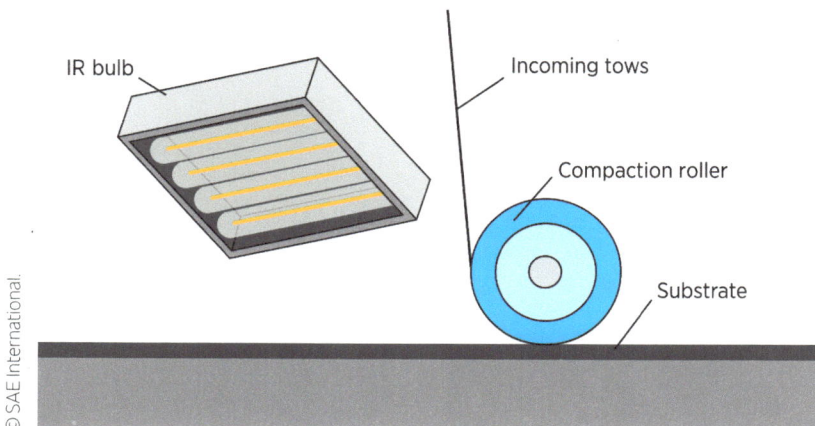

IR bulb

Incoming tows

Compaction roller

Substrate

© SAE International.

Laser heaters are usually employed for thermoplastic layups and have shown themselves to be a better heating option than other heating systems. While laser heaters often exceed temperature limits for thermoset materials, recent systems have demonstrated successful implementation and even additional advantages on thermosets with a suitable heating range. The first concept of laser heaters was demonstrated by Beyeler et al. [2.9] in the 1980s. Lasers have become robust with lower costs, making them a commercially available process. Newer laser systems use a light wavelength that will heat the fibers instead of the matrix preventing damage to the material. A laser system's advantages consist of high energy density, more focused heating, faster processing rates, and better surface finish. Experiments have also shown that laser-assisted AFP has comparatively better interlaminar strength versus placement rate [2.10]. The main disadvantage of these systems is the necessary safety precautions that are required. Typically, laser shielding is required around the AFP cell along with personal protection equipment (PPE) to prevent any reflections from harming personnel. Further, laser heating cannot be used in some materials such as glass fibers since glass fibers do not absorb the laser's energy.

Recent advances in heating systems have made it possible to use heating sources that are more targeted and capable of producing the intended result. An example of such a heating system is Humm3®, developed by Heraeus [2.11]. The mechanism uses light emittance to transfer the heat energy and convert intensity and frequency into a usable value. **Figure 2.18** shows this system's schematic, and **Figure 2.19** shows the system as installed on ADD composites AFP-XS layup head.

Figure 2.18 Humm3 conceptual schematic.

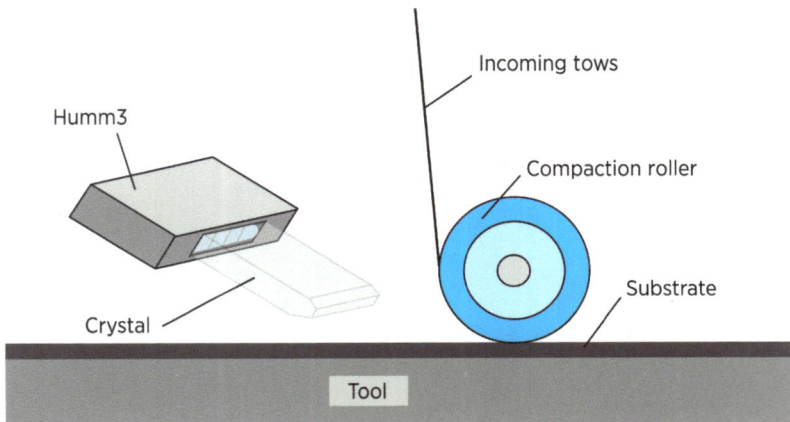

© SAE International.

Figure 2.19 Humm3 on the AFP machine.

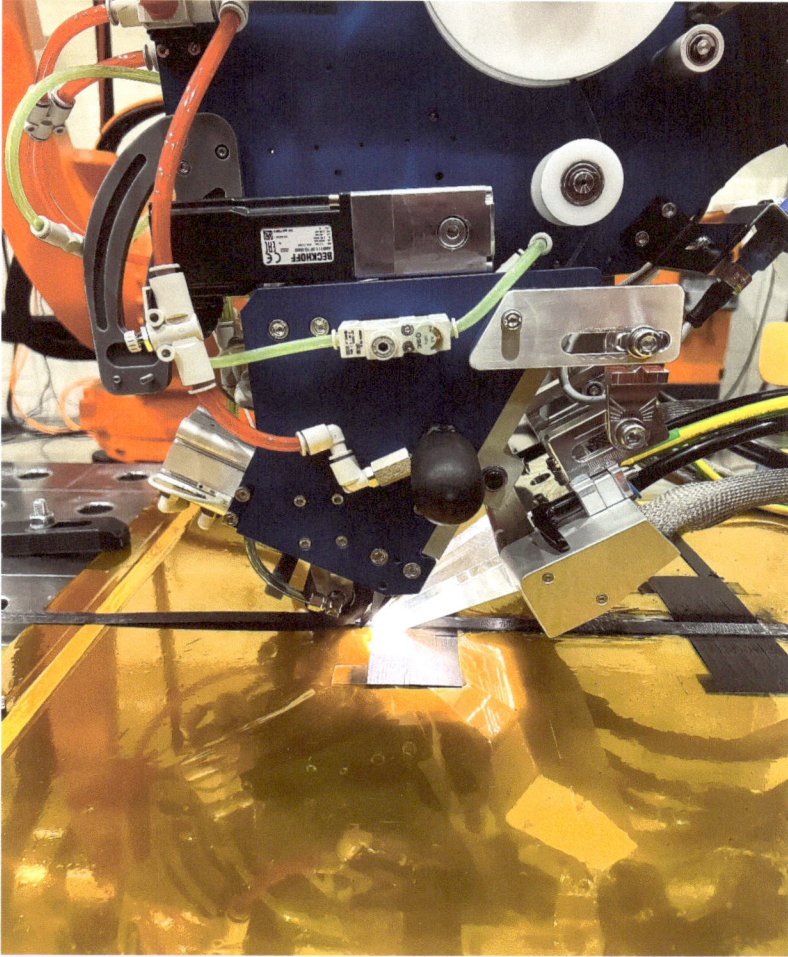

Polymers

Duration: 20 minutes

Description: This video, part of "Composites A-Z: 30 Days of Composites," explores polymers as matrix materials in composites, focusing on thermosets and thermoplastics. It covers polymer properties, applications (e.g., roof rails, tires), and polymerization, highlighting thermoplastic recyclability and thermoset cross-linking, with thermosets being more common in composites.

Scan the QR code below to watch the video.

2.4.
Speed

A very conflicting process parameter is the speed at which the placement occurs. As with every manufacturing facility, production wants to achieve the fastest level of material placement to accelerate production and ensure demand is met. This is sometimes contradictory to both the machine's ability and the quality of obtained layup. This becomes even more crucial once complex shapes are being manufactured as they have several steering locations and/or other parameters to consider. Various layup velocities show alterations in layup quality and required processing parameters [2.12, 2.13]. Lower speeds result in longer thermal exposure, which makes for improved polymer healing up until the applied temperature results in degradation of the material [2.14]. An increase in layup speed will result in less time that the compaction force and temperature are applied to the material, leading to weak cohesive forces [2.15]. The effects of speed on applied temperature and compaction can be summarized with the term "dwell time." Lower speeds lead to higher dwell times and, therefore, longer exposures to heat and compaction pressure. The opposite is true for higher speeds.

Also, process development considers how to appropriately connect the different process parameters. This is then translated into settings that can be achieved simultaneously. Speed can be thought of as an indirect parameter because it may not directly affect the quality of the placement; however, its resulting effects on other process parameters lead to variations in quality. For a better start to a course, there might need to be an increased amount of heat and a slower operation speed until the machine reaches a certain stable period; then, it can have a more uniform process. The same is true for the first ply of a layup. In both cases, it is more difficult to accurately place the material with high quality. At the beginning of a course, this is attributed to the wandering of the tows as they initially are fed out of the AFP head. Once proper adherence and placement are achieved, the tows align properly. For this reason, the as-manufactured ply boundary is often larger than the intended boundary. The excess is then trimmed off during postprocessing. However, this is not always possible as some plies are placed internally to the outer boundary.

The issues seen in the first ply are attributed to the difficulty of getting proper adherence of the incoming material to the tool. The tool often absorbs a lot of the heat from the heating system making the material not adhere as well. Often "tackifier," a combination of acetone and resin from the material or provided separately by the material supplier, is applied to the tool surface to assist with adherence, sometimes in combination with heated tooling. Another case is if the projected path is curvilinear, then the machine will "steer" the tow—not intentionally but forced by the geometry. If the speed slows down for the steered course and heating is still steady, the substrate will burn, and this will negatively affect the quality of the final substrate. This can be alleviated by connecting the heating system to the speed parameter achieving variable temperature application.

2.5.
Tension

Tow tension is fundamental to making sure the tow is in shape while it is being placed. The central idea is that tow tension assists in the placement of tows [2.16]. Excessively high tow tension leads to tow slips due to the tension force overcoming the adherence [2.17]. High tow tension in concave areas of a tool surface can also lead to bridging. The tow tension results in a buildup of stress as the tow is being placed, and once this stress exceeds the adhesive bond, the tows lift and bridge. Additionally, the material aerial weight can heavily influence the needed and appropriate tow tension. In a complex case, roping of tows is completely possible for low aerial weight material. The less aerial weight, the more complexity is needed to find the delicate value that can ensure placement while not operating on extremely low speed/tension combinations.

Tow tension is also crucial when placing steered paths or paths with in-plane curvature. A higher tension force makes the tows less pliable, therefore harder to place at smaller radii. A lower tow tension results in the material being easier to deform prompting better placement of steered paths. Note that the limitation of in-plane radii is a function of material properties as small radii result in out-of-plane buckling and folding of the tows. Adequate tow tension, among the other parameters, can aid in reaching this physical limit without process parameter-induced defects.

2.6.
Compaction

The goal of a roller is to provide compaction and/or consolidation based on the matrix material. The roller is comprised of an interior metal layer with a rubber, typically silicon, outer layer. Compaction pressure is one of many major parameters associated with final part quality. The pressure is the critical parameter to develop intimate contact between plies; however, excessive compaction can lead to material degradation [2.18]. **Figure 2.20** shows a typical roller with and without wrap. The wrap, usually made from a Teflon material, helps keep the material from sticking to the roller. This prevents downtime of the machine by limiting the need to clean the compaction roller of deposited resin.

Figure 2.20 Regular roller with and without wrap.

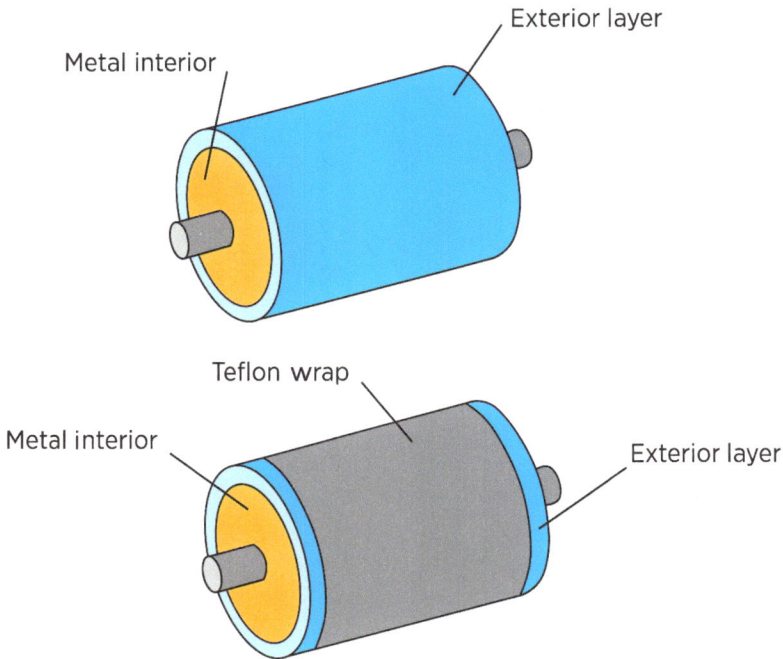

© SAE International.

Figure 2.21 shows one of the recent changes made to rollers. The roller is segmented in a way that will change and deviate based on the tool shape that is being formed. This will help achieve better compression into areas that are typically not accessible with a solid single roller. These are often used with longer rollers since they struggle to achieve adequate consolidation pressure across the entire length, especially for geometrically complex surfaces. Another change is the creation of rollers with the ability to circulate fluid to help with the cooling process. This has become a necessity with the increased exploration of thermo-plastic materials. The higher applied temperatures can quickly degrade the roller. The circulated liquid dissipates some of the heat and increases the life of the compaction roller.

Figure 2.21 Segmented compaction roller.

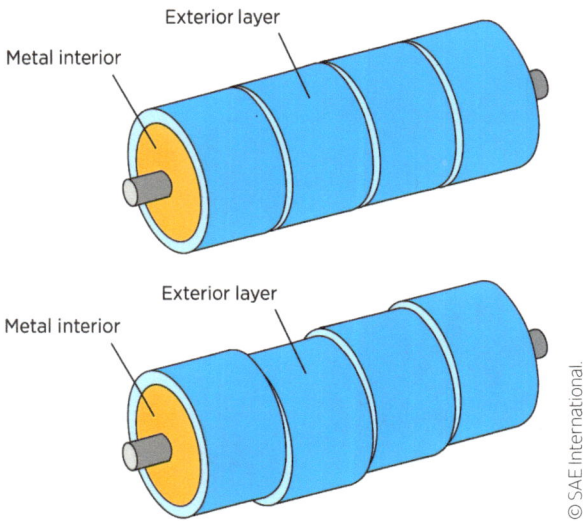

Perforated rollers are also in use in research and industry environ-ments. These rollers have various geometries cut out of the rubber exterior layer. This makes the roller more conformable to the tool surface and is beneficial when manufacturing with complex tool geometries. The perforations in the exterior layer can lead to small variations in applied compaction pressure. Areas directly below a perforation will apply slightly less pressure than those directly below an area without a perforation. However, these variations are minimal and likely not to be detectable in a manufacturing environment.

2.7.
Environmental Monitoring

While we have presented the main process parameters for laying up fibers and the manufacturing of laminates, it is also important to note that the manufacturing environmental conditions must be assessed and monitored for consistent and quality manufacturing. Elements to observe and watch for are as follows:

- Temperature: Maintaining the temperature of the manufacturing environment at $70 \pm 5°F$ is highly desired to have a process in control. Higher temperatures can lead to material, mainly thermosets, adhering to the AFP machine hardware, as well as increased resin and fuzz deposits. Excessively elevated temperatures can also reduce the achievable amount of out time if a creel is not utilized. AFP head cooling is another crucial temperature consideration. For thermoset materials, it is essential to maintain low head temperatures, often below ambient levels, and adjust them based on material tack to prevent fuzz or resin buildup within the head. This becomes particularly critical with IR heaters, where the AFP head tends to warm up during operation, increasing material tack and the risk of defects or FOD.

- Humidity: A humidity range of around 40% is highly desired for optimal laminate manufacturing, although it is not a strict requirement. While there is no direct impact on the final product, maintaining consistent humidity is considered the best practice for achieving manufacturing excellence. It is also important to note that humidity fluctuations can affect material tack, which is undesirable in the manufacturing process.

It is important to highlight that AFP layup should always occur in clean environments with the goal of having parts that are FOD-free. If FOD presents on a laminate, it requires removal or rework that increases machine downtime. In research and experimental studies, the environment becomes even more critical as we need to compare the experimental data in between experiments. If the production is taking place at completely different environmental ranges, it will influence the data and the interpretation of the experiment. Therefore, we often encounter research laboratories with tighter tolerances than production environments. Better clean room classification results in higher monitoring of the quality. AFP rooms are seen operational in ISO 7 environments.

Did You Know?

Across various manufacturing disciplines, process parameters are a prevalent attribute and often share similarities. However, certain manufacturing processes may have distinct elements within the same attribute. For instance, in machining, speed is separated into two distinct parameters: cutting speed, which relates to rotation, and feed rate, which is similar to AFP and additive manufacturing in general. Determining the optimal process parameter conditions for each manufacturing method typically requires trial and error, particularly when physical modeling does not yield accurate predictions, which is frequently the case.

2.8.
Conclusion

Identification of appropriate values for process parameters is determined through a qualitative analysis of defects at various fiber orientation angle placements. Ideally, the placement should entail a minimal number of defects, while the part is fabricated with a high feed rate. Certain types of defects can be correlated or attributed to a specific process parameter, which can be consequently adjusted. All AFP defects are detailed in subsequent chapters. There are other elements that influence the appropriate design of experiment such as when we start tailoring equipment and elements like redirect rollers (**Figure 2.22**) and cutters (**Figure 2.23**).

Figure 2.22 Redirect rollers.

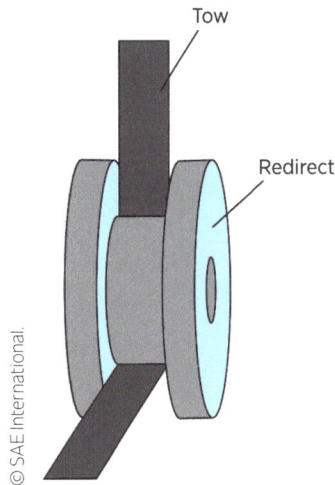

Figure 2.23 Cutters.

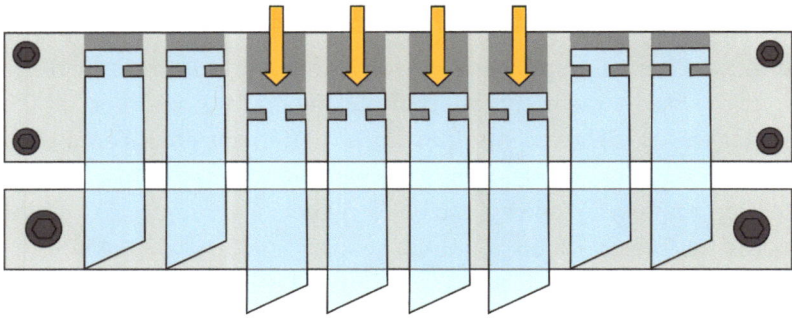

© SAE International.

This chapter is best served ending with a methodology for process parameters investigation and the appropriate design of experiment. A sample approach is detailed in Harik et al. [2.19], and **Table 2.2** shows some of these results.

Table 2.2 Process parameters during manufacturing trials [2.19].

Trial	Fiber orientation angle [°]	Courses	Heater output [%]	Compaction [N (lb)]	Feed rate [%]	Tension [N (lb)]
1	+45	1–8	200	222.41 (50)	40	4.45 (1)
2.1	0	1–3	200	222.41 (50)	40	4.45 (1)
2.2	0	4–6	150	222.41 (50)	60	4.45 (1)
2.2	0	7–8	150	222.41 (50)	70	4.45 (1)

R. Harik, J. Halbritter, D. Jegley, R. Grenoble and B. Mason, "Automated Fiber Placement of Composite Wind Tunnel Blades: Process Planning and Manufacturing," in SAMPE Conference & Exhibition, Charlotte, North Carolina, US, 20 – 23 May 2019. Reprinted by permission from the Society for the Advancement of Material and Process Engineering (SAMPE).

If operating in a clean room and conditioned environment, where temperature and humidity are kept within certain limits, it will result in the factors of our design of experiment to:

- Speed
- Temperature
- Tension
- Compaction

The next stage would be to determine the levels of each one of those parameters that can be used. Example ranges are provided below; however, appropriate choices should be made based on knowledge of the AFP machine and the material system. These examples can be taken as generic estimates.

- Speed: low level of 500 ipm; high level of 2000 ipm
- Temperature: low level of 50%; high level of 100%
- Tension: low level of 0.5 lb; high level of 1.0 lb
- Compaction: low level of 50 lb; high level of 150 lb

With the ranges of parameters defined, a test matrix is developed to evaluate each parameter and its combined effect on the quality of the produced part. This test matrix should also include manufacturing at multiple ply angles as the direction the AFP is traversing the part can influence the necessary process parameters, especially when considering complex geometries. The interval and total number of combinations of parameters are dependent on the availability of machine time and the desired cost of the experiment. A full test campaign should adequately assess the influence of each parameter alone and in combination with the other parameters.

The design of experiments must also define the quality metric that will define the successfulness of a parameter selection. For instance, in a solely temperature-based experiment, the quality metric could be whether the material has adhered to the surface or not. Therefore, the optimal temperature range will be the one that achieves complete adherence for all material placements. However, in combined parameter experiments, it is best to consider a range of defects. Here, one would identify what is a parameter-related defect and what is related to external factors such as machine program or material deficiencies. For example, defects such as gaps, overlaps, tow wandering, bridging, and wrinkles (all of which are described in the following chapter) can be directly affected by process parameters. Defects such as splices are material-dependent, while things such as twists are mostly an effect of machine setup.

With the parameter test matrix defined and the quality metric established, manufacturing trials begin. Note that appropriate toolpath planning is also necessary to avoid path-related defects. This will be the topic of a later chapter. Assessment of manufacturing trials could look something like the image in **Figure 2.24**. Various parameters have been used in each course, and defects have been labeled. This process is continued until appropriate manufacturing parameters are determined. It is also possible to quantify the quality metric and build a more numeric analysis of the manufacturing trials. For instance, a total number of defects could be counted, or total area of a certain defect can be used if an inspection system is available. Such an approach also allows for models to be built that can predict quality based on process parameters. Often a response surface is the chosen method to evaluate the effect of multiple parameters on a chosen objective.

Figure 2.24 Example quality assessment of process parameter trials [2.19].

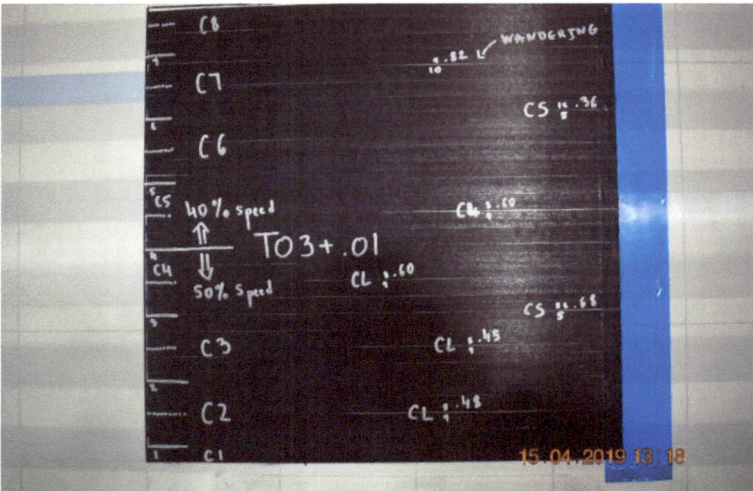

R. Harik, A. Lovejoy, D. Jegley and C. Yokan, "Thin-ply: Exploration and Manufacturing with Automated Fiber Placement," in CAMX Conference & Exhibition, Orlando, Florida, 21-24 September 2020. Reprinted by permission from the Society for the Advancement of Material and Process Engineering (SAMPE).

References

2.1. Harik, R., Halbritter, J.A., Jegley, D., Grenoble, R. et al., "Automated Fiber Placement of Composite Wind Tunnel Blades: Process Planning and Manufacturing," in *International SAMPE Technical Conference*, vol. 2019, Charlotte, May 2019, 1-16, doi:10.33599/nasampe/s.19.1538.

2.2. Jeffries, K., "Enhanced Robotic Automated Fiber Placement with Accurate Robot Technology and Modular Fiber Placement Head," *SAE Int. J. Aerosp.* 6, no. 2 (2013): 774-779, doi:https://doi.org/10.4271/2013-01-2290.

2.3. Flynn, R., Rudberg, T., and Stamen, J., "Automated Fiber Placement Machine Developments: Modular Heads, Tool Point Programming and Volumetric Compensation Bring New Flexibility in a Scalable AFP Cell," in *SME Composites Manufacturing Conference*, Dayton, 2011.

2.4. Flynn, R., Nielson, J., and Rudberg, T., "Production Implementation of Multiple Machine, High Speed Fiber Placement for Large Structures," *SAE Int. J. Aerosp.* 3, no. 1 (2010): 216-223, doi:https://doi.org/10.4271/2010-01-1877.

2.5. Mantell, S. and Springer, G., "Manufacturing Process Models for Thermoplastic Composites," *Journal of Composite Materials* 26, no. 16 (1992): 2348-2377.

2.6. Qureshi, Z., Swait, T., Scaife, R., and El-Dessouky, H., "In Situ Consolidation of Thermoplastic Prepreg Tape Using Automated Tape Placement Technology: Potential and Possibilities," *Composites Part B: Engineering* 66 (2014): 255-267.

2.7. Rizzolo, R. and Walczyk, D., "Ultrasonic Consolidation of Thermoplastic Composite Prepreg for Automated Fiber Placement," *Journal of Thermoplastic Composite Materials* 29, no. 11 (2016): 1480-1497.

2.8. Greenberg, B.L., "Design and Control of an Arrayed Infrared (MAT-IR) Heater for Accurate Heating Control during Automated Fiber Placement," Doctoral dissertation, 2020, Retrieved from https://scholarcommons.sc.edu/etd/5663.

2.9. Beyeler, E., Phillips, W., and Güçeri, S.I., "Experimental Investigation of Laser-Assisted Thermoplastic Tape Consolidation," *Journal of Thermoplastic Composite Materials* 1, no. 1 (1988): 107-121, doi:10.1177/089270578800100109.

2.10. Stokes-Griffin, C.M. and Compston, P., "The Effect of Processing Temperature and Placement Rate on the Short Beam Strength of Carbon Fibre-PEEK Manufactured Using a Laser Tape Placement Process," *Compos Part A Appl Sci Manuf* 78 (2015): 274-283, doi:10.1016/j.compositesa.2015.08.008.

2.11. Heraeus, "humm3TM - Intelligent Heat for Automated Fibre Placement (AFP)," 2021, Retrieved from https://www.heraeus.com/en/hng/products_and_solutions/arc_and_flash_lamps/humm3/humm3.html#tabs-115105-1.

2.12. Chen, J., Chen-Keat, T., Hojjati, M., Vallee, A. et al., "Impact of Layup Rate on the Quality of Fiber Steering/Cut-Restart in Automated Fiber Placement Preocesses," *Science and Engineering of Composite Materials* 22, no. 2 (2015): 165-173.

2.13. Engelhardt, R., Irmanputra, R., Brath, K., Aufenanger, N. et al., "Thermoset Prepreg Compaction during Automated Fiber Placement and Vacuum Debulking," *Procedia CIRP* 85 (2019): 153-158.

2.14. Khan, M.A., Mitschang, P., and Schledjewski, R., "Identification of Some Optimal Parameters to Achieve Higher Laminate Quality through Tape Placement Process," *Advances in Polymer Technology* 29, no. 2 (2010): 98-111.

2.15. Belhaj, M., Deleglise, M., Comas-Cardona, S., Demouveau, H. et al., "Dry Fiber Automated Placement of Carbon Fibrous Preforms," *Composites Part B: Engineering* 50 (2013): 107-111.

2.16. Zhang, W., Liu, F., Lv, Y., and Ding, X., "Modelling and Layout Design for an Automated Fibre Placement Mechanism," *Mechanism and Machine Theory* 144 (2020): 103651.

2.17. Assadi, M. and Field, T., "AFP Processing of Dry Fiber Carbon Materials (DFP) for Improved Rates and Reliability," *SAE Int. J. Adv. & Curr. Prac. in Mobility* 2, no. 3 (2020): 1196-1201, doi:https://doi.org/10.4271/2020-01-0030.

2.18. Zacchia, T.T., Shadmehri, F., Fortin-Simpson, J., and Hoa, S.V., "Design of Hard Compaction Rollers for Automated Fiber Placement on Complex Mandrel Geometries," in *Proceedings of The Canadian Society for Mechanical Engineering International Congress 2018*, Toronto, ON, Canada, 2018.

2.19. Harik, R., Lovejoy, A., Yokan, C., and Jegley, D., "Thin-Ply: Exploration and Manufacturing with Automated Fiber Placement," in *CAMX 2020*, Virtual Experience, 2020.

Automated Fiber Placement equipment at the National Institute for Aviation Research in Wichita, Kansas.

Chapter 03

AFP Defects

AFP manufacturing can result in many defects during the layup process that often require manual corrective action to produce a part with acceptable quality. These defects are the main limitation of the technology and can be hard to categorize or define in many situations. This chapter provides a thorough definition and classification of all AFP defects. This effort constitutes a comprehensive and extensive library relevant to AFP defects and was initially published in [3.1] and [3.2]. The defects selected and defined in this work are based on understanding and experience from the manufacture and research of advanced composite structures. The results are identity (ID) cards for each defect type, intended to provide researchers and the manufacturing industry with a clear understanding of the (1) cause, (2) anticipation, (3) existence, (4) significance, and (5) progression of the defined AFP defects.

3.1.
Viewpoint Modeling

To fully understand AFP defects, a complete understanding of the source and how the part geometry might influence the defect formation must be reached. The current work begins with a thorough categorization of AFP defects to develop an understanding of the importance of defects from five different perspectives: the cause, anticipation, existence, significance, and progression. Each of these categories has a specific perspective of defining what is considered a defect:

- The **cause category** investigates the core cause or causes of a defect. What we present in this section is believed to be the most likely cause, although it should not be interpreted as definitely conclusive.

- The **anticipation category** reflects the perspective of the process planner, investigating the predictability of defects occurring under specific parameters. This category will show possible parameter changes to avoid the defect.

- The **existence category** defines the defect from an inspection point of view and what the defect visually looks like. This category will show whether a defect is better suited for either manual, semi-automated, or fully automated inspection methods. Additionally, this category should be expanded to include observations post-cure. This is particularly important in the case of thermoset materials, where defects can be significantly altered during the curing process.

- The **significance category** investigates the level of significance a particular defect can have on a part if it occurs. In addition, this section investigates what the defects can lead to if left unresolved from a design perspective.

- The **progression category** fully investigates if the defect will progress in the part under service loads if nothing is done to fix the affected site.

Application

In materials science, a general rule of mixtures is a weighted mean used to predict various properties of a composite material. This concept can be used for composites as well by combining the properties of both fiber and resin.

In general, two equations can be used to represent a composite elastic modulus as shown below. The first is considered an upper bound, while the second is considered a lower bound. Typically, the final value is somewhere in between

$$E_c = fE_f + \left(1 - f\right)E_m$$

$$E_c = \left(\frac{f}{E_f} + \frac{1-f}{E_m}\right)^{-1}$$

In these equations, E_f is the modulus of the fiber, E_m is the modulus of the matrix, E_c is the modulus of the composite, and f is the volume fraction of fibers.

Using the above equations, let us examine how the modulus of the composite changes as a function of the fiber volume fraction. Consider a matrix with a modulus of 5 GPa and a fiber with a modulus of 5 GPa, and a fiber with a modulus of 220 GPa. Note that these are arbitrary values and do not signify a specific type of either constituent. Using the equations above, we can plot the modulus of the composite as shown in the figure below. The actual value of the modulus often falls within these two lines. Much effort can be found in literature to develop similar equations that more accurately define the exact behavior of a composite material.

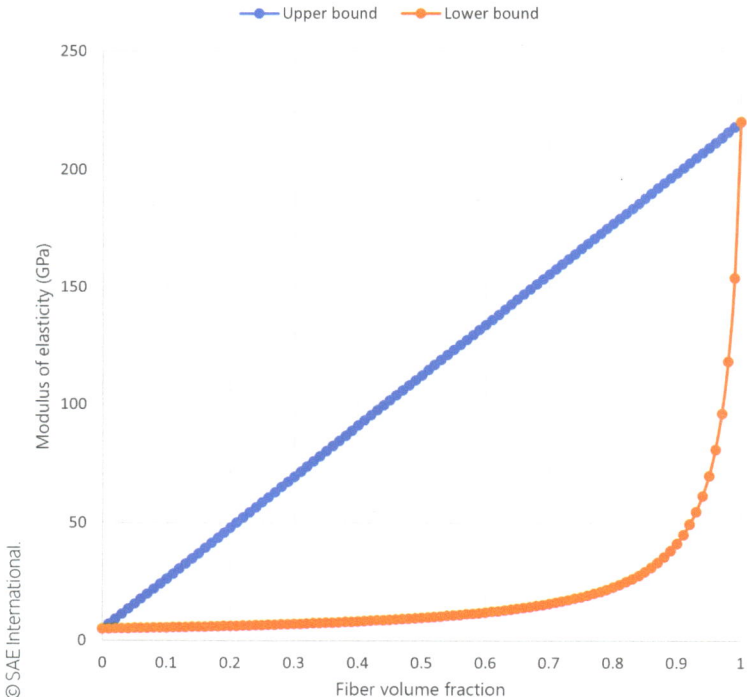

3.2.
Defect ID Cards

We define 14 of the most common defect types, providing both a graphical representation and a discussion on their relevance from four different perspectives. For most defects, we propose a computer-aided design (CAD) representation and a picture from the actual manufacturing trials.

3.2.1.
Gap/Overlap

A gap is when two adjacent tows are not perfectly laid up next to each other, and there is a gap between the two. An overlap is similar, but the two adjacent tows are overlapping onto each other. The most common cause of gaps and overlaps is steering during layup since the tows in a course will not fit together perfectly, especially when adopting a constant angle coverage strategy. However, gaps and overlaps can naturally occur outside of steering if laying up over a complex 3D tool surface. This is attributed to the impossibility of uniformly covering a complex surface with constant width tows (**Figures 3.1**, **3.2** and **Table 3.1**).

Figure 3.1 Rendering of a gap/overlap model.

© SAE International.

Figure 3.2 Image of an actual gap defect.

Table 3.1 Gap/overlap characteristics.

Anticipation	Purposeful steering creates well-anticipated gaps or overlaps. Doubly curved parts will produce either gaps or overlaps since courses will not have perfect coverage of the surface. Based on the exact surface geometry and the path of the machine head, the amount of gap/overlap can be computed.
Existence	Gaps and overlaps are easy to detect since they will be visible by the lack of or addition of material when compared to a regular course. As successive plies are laid up over overlaps, consistently at the same location at each layer, significant thickness buildup will be visible.
Significance	A gap and/or overlap may become a site for failure initiation under loads. Gaps would create resin-rich regions for crack growth. Overlaps create undulations in the fiber that can lead to compressive failures. They may also become a site for wrinkling in the layers placed over them in the succeeding layer.
Progression	Progressive failure events under static and fatigue loads are a current topic of research by the authors. It may be too speculative to describe micromechanics of failures at this time.

3.2.2.
Pucker

A pucker is initiated at the inside radius of a steered tow. This results in the tow lifting from the tool surface either partially or across the entire tow width, forming an arch of excess material that is not adhered to the underlying substrate material. Puckered tows are caused by excess feeding of a tow that gradually accumulates ahead of the compaction roller and, at some point, emerges in the part surface. If placement is over a compliant surface, with the force of the compaction roller, longer tows may be deposited that can form the pucker after the surface springs back to its original shape (**Figure 3.3** and **Table 3.2**).

Figure 3.3 Rendering of a pucker model.

© SAE International

Table 3.2 Pucker characteristics.

Anticipation	Can be prevented through appropriate towpath planning. Thickness buildup in concave shapes may cause shortening of the surface length, and the tow length paid-out by the machine head may need to be shortened to compensate for the reduced length.
Existence	Small puckers may be difficult to detect visually. A profilometry sensor-based detection system can be utilized to identify the puckered tow.
Significance	A puckered tow is typically flattened by successive layers placed over them and by debulking. However, if the puckers are not properly compacted, it may result in a significant loss of strength.
Progression	Like most defect types, the progressive failure of puckers is not investigated. However, if the pucker is not flattened by placement of the next layer, delamination growth should be expected between the folds under cyclic loads that can propagate to become delamination between the layers.

© SAE International

3.2.3.
Wrinkle

A wrinkle is typically indicated by a wavy pattern of puckering along the edge of a tow when it is steered through a non-geodesic path over a complex (potentially doubly curved) surface or following a steered path on a flat surface. These types of defects occur on the inner radius and remain out of plane after compaction and curing. Wrinkles are often caused by placing tows at small steering radii, which can lead to excessive differential length between the two edges of the projection of the tow on the part surface. The two edges of a tow delivered from the machine head are equal in length; hence, part of the excessive differential length presents as puckers and/or wrinkles.

A tow path-based model of wrinkling during the AFP process was developed at USC McNAIR and presented in *The Composites and Advanced Materials Expo* [3.3]. An investigation of wrinkling within an arbitrary path for a composite tow constructed using the AFP process was presented. Governing equations and assumptions for a basic zeroth-order model were derived based on geometric considerations only, neglecting the viscoelastic properties of the material, and formulated for an arbitrary curve on a general 3D surface (**Figure 3.4** and **Table 3.3**).

Figure 3.4 Rendering of a wrinkle model.

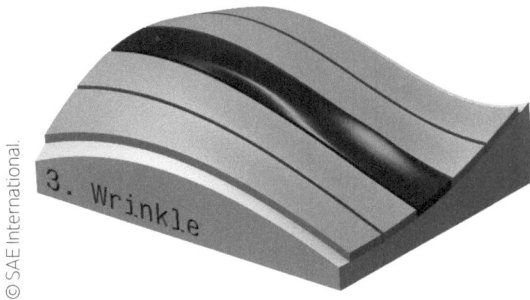

© SAE International.

Table 3.3 Wrinkle characteristics.

Anticipation	The steering radius definition and the complexity of the tool surface being laid up on are the main ways to anticipate a wrinkle. Tow path definition during the design phase can have a strong influence on wrinkling behavior. Process parameters and tow material properties are also influential.
Existence	Can be detected either visually or using automated inspection systems. Can be difficult to distinguish from puckers as the tow is overhanging along its orientation.
Significance	Wrinkled tows covered by layers that are laid on top, forcing them to flatten during which in-plane fiber waviness or folded fibers may be caused. Can lead to gaps and folded tows, potentially resulting in reduced strength.
Progression	Progressive damage of in-plane waviness and the folding are the subject of ongoing research.

© SAE International.

AFP Defects

Duration: 24 minutes

Description: This video, part of "Composites A-Z: 30 Days of Composites," examines defects in automated fiber placement for large structures. It details gaps/overlaps, puckers, wrinkles, and bridging, explaining causes, detection difficulties, and effects like resin-rich areas or reduced strength.

Scan the QR code below to watch the video.

3.2.4.
Bridging

A bridged tow does not fully adhere to the concave surface (female tool portion), a re-entrant corner, or a ramp-up area over which the tows are laid. This results in a gap between the radius of the concave tool surface and the tow. The main causes of a bridged tow are too much tension on the tow, which will force the tow to lift up, or insufficient tack adhesion to the surface being laid up on due to the roller not providing full contact with the substrate material (**Figures 3.5**, **3.6** and **Table 3.4**).

Figure 3.5 Rendering of a bridging model.

© SAE International.

Figure 3.6 Image of an actual bridging defect.

© SAE International.

Table 3.4 Bridging characteristics.

Anticipation	The main way to anticipate this defect is to ensure that the roller has the best contact coverage possible when going over a complex tool surface, especially concave portions. Overfeeding of tow may eliminate bridging in re-entrant corners and ramps.
Existence	Bridging is often readily identified both visually and by automated inspection systems, as the tow in question will be raised about the concave portion of the tool being laid up on.
Significance	Successive passes of the roller to place additional layers with different orientations or a de-bulking step with vacuum may push the bridged tow to re-adhere to the substrate. However, the bridging could leave resin-rich areas at best or delamination at worst.
Progression	Significant consideration should be given to these sites as potential progressive growth of delamination.

© SAE International.

3.2.5.
Boundary Coverage

A boundary gap/overlap occurs when the material cannot perfectly meet up with the edge of a part when laying up at off-axis orientations such as ±45° in rectangular parts. Since the tows do not meet up perfectly with the edge, this will result in either an excess of material along that edge or a shortage between the tow end and the boundary edge. This can be at the boundary of any coverage zone, be it internal to the part inducing ply drop-offs or at the external boundary (**Figures 3.7**, **3.8** and **Table 3.5**).

Figure 3.7 Rendering of a boundary coverage model.

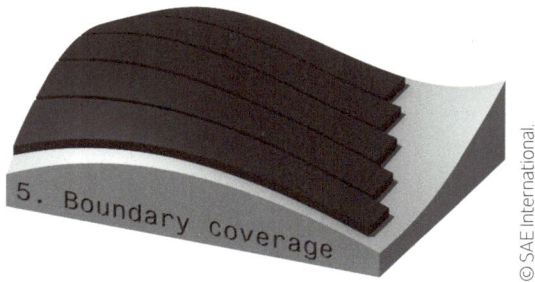

© SAE International.

Figure 3.8 Image of an actual boundary coverage defect.

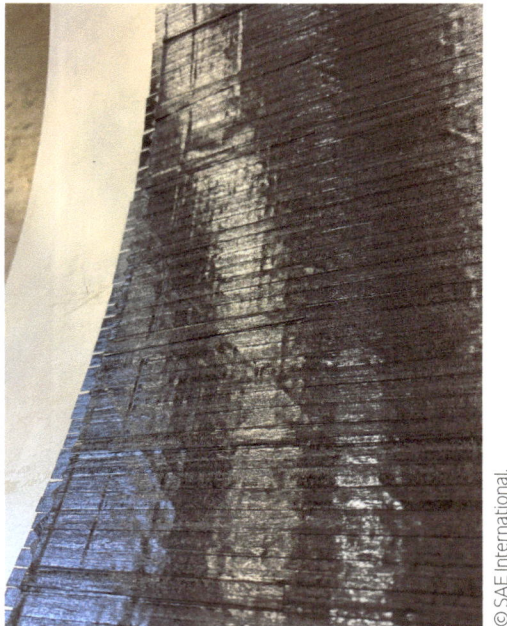

© SAE International.

Table 3.5 Boundary coverage characteristics.

Anticipation	The ways to anticipate this defect are in defining the percentage of boundary gap and overlap and observing the part's geometry in relation to the ply angles.
Existence	These defects are clearly visible on the boundary of any variable angle laminates and will be visible post-cure.
Significance	A boundary gap and/or overlap can have an effect on the shape of the part since the course will not line up with the desired geometry. If the edges are trimmed to ensure accuracy, the part may also become more likely to fail in those spots.
Progression	Not investigated by the authors currently.

Did You Know?

The effect of the AFP defects presented throughout this chapter is a critical aspect that must be considered, especially when undergoing DFM for a structure. Defects can affect a laminate in terms of strength by up to 13%. Much of this effect is attributed to fiber waviness, which can negatively impact the structures. However, some defects can, counterintuitively, positively affect the strength of a laminate. Still, a defect that gives an improvement in one mechanical property also penalizes another property. Therefore, for real parts with multidirectional loads, it is difficult to gauge the advantages of these increases. The table below summarizes experimental results of key defects on various mechanical properties of a laminate.

The influence of these defects, among others, must often be characterized by a specific material or use case. During a DFM effort, the effect of defects must be integrated into the design to accurately predict the performance of the structure. Also, not just the occurrence but the distribution must be considered throughout a laminate. Defects that accumulate in a given area of a structure will more significantly affect the structure than those that are randomly distributed. Effort to position defects in a less structurally critical area of a laminate and prevent defect stacking should be undertaken during planning of AFP manufacturing.

3.2.6.
Angle Deviation

Angle deviation is when the angle of the as-manufactured layup deviates from the as-designed one. Angle deviation can be caused by incorrect roller coverage or small radius steering as the tow may move after being steered. Tool surface geometry can also cause angle deviation depending on the strategy used during path planning (**Figure 3.9** and **Table 3.6**).

Figure 3.9 Rendering of an angle deviation model.

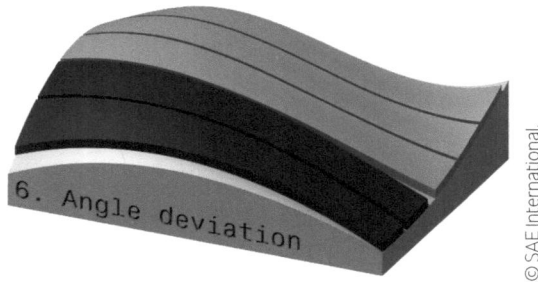

6. Angle deviation

© SAE International.

Table 3.6 Angle deviation characteristics.

Anticipation	The main way to anticipate angle deviation is through defining the steering radius for any required steering throughout the layup, since a smaller radius can cause angle deviation.
Existence	Angle deviation is observed by visual inspection, but it requires further processing and comparison with the as-designed angles. Inspection systems alone cannot confirm the angle deviation without a benchmark for comparison.
Significance	Angle deviation can cause overlap on portions of the ply when a course is laid up on top of the deviated tows. This can lead to an undesired shape in the laminate and can be a cause of failure due to resin-rich areas on the counter side. This is a similar effect to overlap and gaps.
Progression	Can lead to delamination in the resin-rich areas due to improper course coverage.

© SAE International.

3.2.7.
Fold

This defect occurs when the tow folds in the transverse direction onto itself, creating a gap in the surface coverage and doubling the tow thickness over the folded part. An extension (and probably the worst-case scenario) of the folding could be rolling (or completely twisting) of the tow to become "rope"-like. Tensioner errors, lack of or too much tension, could increase the propensity of the tow to fold. Long unsupported/complex tow paths from the spools to the head can also result in folding. In a steered/curved tow path, the outer segment of the tow may fold toward the inner side after the compaction roller nip point due to tension on the outer edge of the tow and improper tack adhesion (**Figure 3.10** and **Table 3.7**).

Figure 3.10 Rendering of a fold model.

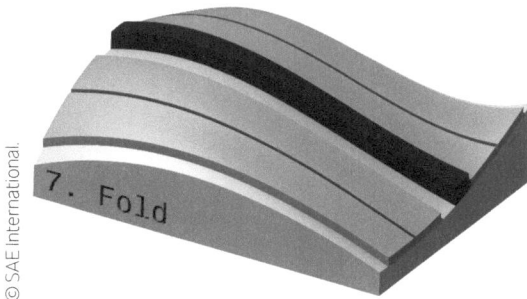

Table 3.7 Fold characteristics.

Anticipation	The quality of the slit tape or tow has a large influence on the folding defect. Machine type and machine calibration (health) would have an impact. Design has some influence if the tow path is steered. Process parameters (speed and temperature) influence the tack adhesion.
Existence	Can be detected either visually or via automated inspection systems.
Significance	One of the more serious defect types for cured laminates is due to increased thickness right next to a reduced thickness region. This results in substantial influence on local fiber volume fraction variation and creation of resin-rich areas for failure initiation.
Progression	Sites for delamination initiation following the crack growth from the resign-rich defect sites.

© SAE International.

3.2.8.
Twist

For this type of AFP defect, the tow is rolled axially 180° onto itself and then flattened by the compaction roller. Depending on the length over which the twisting occurs, the shape may be like a bowtie, with bunching of the fibers and increased thickness at the center. For long twists, the sides are simply folded. Twisted tows could be initiated by folding, in which the fold grows and completes a full turn rather than unfolding (a folded tow could be considered an incomplete twist). Friction between guide holes along a long/complex tow path and a tacky tow may cause twisting due to head rotation during bidirectional layups (**Figure 3.11** and **Table 3.8**).

Figure 3.11 Rendering of a twist model.

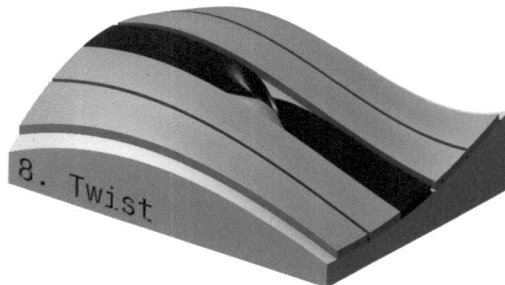

© SAE International.

Table 3.8 Twist characteristics.

Anticipation	The rotation of the machine head, the geometry of the part, and the tows not being properly fed into the machine can all contribute to having a twisted tow during layup. The geometry of the part contributes since a head rotation may be necessary on some portions of the part surface.
Existence	Can be detected either visually or via automated inspection systems. Machine learning algorithms can be useful in classifying the defect as a twist.
Significance	Like folded tows, twisting results in a portion of the part surface not covered with fiber, especially for long twists, and parts with increased thickness. A twist may be considered to be more damaging than a fold, as structural load may cause scissoring deformation.
Progression	Besides being a source for cracking and delamination, severe deviation of the fiber paths within the tow from being straight will cause kinking failure of tows under compressive loads.

© SAE International.

3.2.9.
Wandering Tow

A wandering tow is when the portion of the tow between the roller and the cutter wanders from the original fiber path after being cut. Similar to tow "angle deviation," wandering tows are more attributable to having an unsupported portion of the tow between the compaction roller and the tow cutter, and therefore, the angle deviation will only be of the dimension of this unsupported tow length (**Figure 3.12** and **Table 3.9**).

Figure 3.12 Rendering of a wandering tow model.

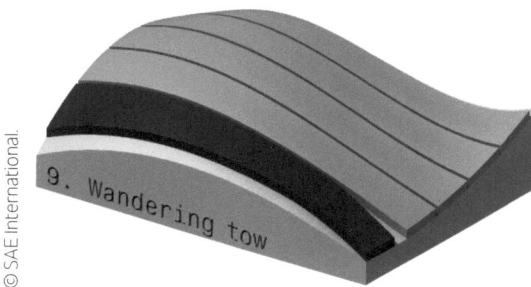

9. Wandering tow

© SAE International.

Table 3.9 Wandering tow characteristics.

Anticipation	The main approach to anticipate, and thereby prevent, a wandering tow would be in ensuring that any steering does not have too small of a radius or that the roller coverage is maximum to ensure proper adhesion.
Existence	Wandering tows can be visually observed since they are typically located at the ends of a course.
Significance	Wandering tows can lead to a gap/overlap between tows which can result in a resin-rich area and ultimately a higher chance of failure.
Progression	Can lead to failure within the laminate due to any gaps created in the ply.

© SAE International.

3.2.10.
Loose Tow

A loose tow generally refers to a section of a tow (or tows) that the machine head attempts to place on a part without having complete and precise control over where it is actually placed, causing the tow to meander. A tow is completely loose when the length of a tow is shorter than the length between the cutters and the compaction roller that controls the tow's final position. This distance is often referred to as the minimum cut length in AFP. In this case, the tow is free to land in an arbitrary position. If at the end of a course the fiber path is still steered, the section of the tow before the compaction roller may not follow the defined steered path (**Figures 3.13**, **3.14** and **Table 3.10**).

Figure 3.13 Rendering of a loose tow model.

© SAE International.

Figure 3.14 Image of an actual loose tow defect.

© SAE International.

Table 3.10 Loose tow characteristics.

Anticipation	Angle plies at rectangular corners or other geometric features that require short tows will need to be forced to have longer tow placement, resulting in features called "dog ears" or "bat wings." Tow paths for parts that need steering near edges can also be extended beyond their boundaries to eliminate meandering of tows.
Existence	If the loose tow results in a significant gap in the laminate, or a completely missing tow, then it can be detected visually or by automated vision systems.
Significance	Missing tows are already discussed with their own ID card. If the loose tows are because of steering, then its consequences must be accounted for by using tools appropriate for it (if/when they exist). If it causes an unanticipated gap or overlap, refer to appropriate ID card.
Progression	Described in missing tow and tows with gaps and overlaps.

© SAE International.

3.2.11.
Missing Tow

This defect typically occurs when an entire tow does not correctly adhere and falls off the surface or is not successfully fed onto a surface from the spools. The resulting missing tow is very similar to a gap and, in fact, can be considered as a gap with a size equal to a tow width. Missing tows are caused by either discontinued material feeding into the machine head or layup of a tow with insufficient tack adhesion (**Figures 3.15**, **3.16** and **Table 3.11**).

Figure 3.15 Rendering of a missing tow model.

© SAE International.

Figure 3.16 Image of an actual missing tow defect.

© SAE International.

Table 3.11 Missing tow characteristics.

Anticipation	This type of defect is not related to any designed features. Ensuring proper splices and full material spools will eliminate accidental missing tows. On complex surfaces, providing sufficient compaction pressure and ensuring sufficient material tack with proper temperature will preclude long-bridged tows that may fall off the surface.
Existence	The gaps created by missing tows are easy to detect either visually or through automated hardware.
Significance	Like a gap, missing tows will cause local thickness variation and potential resin-rich pockets in the layup that can serve as a failure initiation point.
Progression	This defect is a potential site for progressive delamination failure with the adjacent layers.

© SAE International.

Composites with ElectroImpact

Duration: 48 minutes

Description: This video, part of "Composites A-Z: 30 Days of Composites," highlights Electroimpact's AFP capabilities, showcasing their high-speed, precision layup systems for aerospace structures. It emphasizes modular machine designs, multi-tow heads, and path planning. Electroimpact's AFP technology delivers efficient, repeatable composite manufacturing tailored for large-scale aircraft components and next-generation production needs.

Scan the QR code below to watch the video.

3.2.12.
Splice

A splice is when two tows are joined together by the material or slitting supplier end to end in a spool by overlapping 1 to 3 in. over each other and tacking them together. This results in a portion of the spool that is thicker than the rest and is usually marked by white dashes for detection. Theoretically, carbon fibers can be drawn infinitely long. However, most AFP pre-impregnated tows are slit tape that are cut from a roll of finite-length unidirectional tape. These slit tapes are spliced and spooled based on customer specifications (**Figures 3.17**, **3.18** and **Table 3.12**).

Figure 3.17 Rendering of a splice model.

© SAE International.

Figure 3.18 Image of an actual splice defect.

© SAE International.

Table 3.12 Splice characteristics.

Anticipation	Monitoring and keeping track of spool length for the splice locations with respect to the part size may completely eliminate the spliced tow from the part.
Existence	Splices are difficult to detect visually if not marked. The thickness increase over the splice allows detection with a detection system.
Significance	Thickness change over a small area may be insignificant for stiffness change. This site may become a location for failure initiation especially under compressive loads.
Progression	Splices are possible sites for fiber kinking progression under compressive loads.

© SAE International.

3.2.13.
Position Error

A position error is when a tow is placed in the wrong location in reference to the beginning or end of a course. This results in a tow that is misaligned with the rest of the tows in the boundary. Main causes of this defect are either obstruction of the tow during feeding (such as building up of fuzz in one of the guide chutes of the machine head) or incorrect machine reference points with respect to the part for a particular course. Sometimes, they are due to machine control issues and auto-tuning requirements (**Figures 3.19**, **3.20** and **Table 3.13**).

Figure 3.19 Rendering of a position error model.

© SAE International.

Figure 3.20 Image of an actual position error defect.

© SAE International.

Table 3.13 Position error characteristics.

Anticipation	Position errors are arbitrary, and there is no way to anticipate. However, ensuring that there is no material buildup in the head of the machine that can cause resistance during layup, and monitoring the accuracy of the layup simulation will reduce the possibility of occurrence.
Existence	Detected visually or by automated techniques.
Significance	Similar to the tow gap close to the part boundary, their influence is expected to be more pronounced compared to regular defects, due to edge-effect failures observed in multilayered composites.
Progression	Unknown, but can be expected to be like missing tows or gaps.

© SAE International.

Did You Know?

The interior structure of an aircraft wing is more complex than expected. There are critical members such as the spars, ribs, and skin. The spar is often the main structural member of the wing, running roughly spanwise at right angles to the fuselage. The spar carries flight loads and weight of the wings while on the ground. The ribs run along the wing's chord length, providing support and shape to the structure. The skin of the wing is the outer surface and transmits in-plane shear loads into the surrounding structure and gives the wing its aerodynamic shape. AFP is often used to manufacture spars and skins in state-of-the-art composite aircraft.

There are other nomenclatures such as leading edge, trailing edge, root, and tip. The leading edge is the front of the wing, while the trailing edge is the back. The root is the portion of the wing that is attached to the fuselage, and the tip is the outermost end of the wing. A simplistic diagram of a wing can be seen in the following figure.

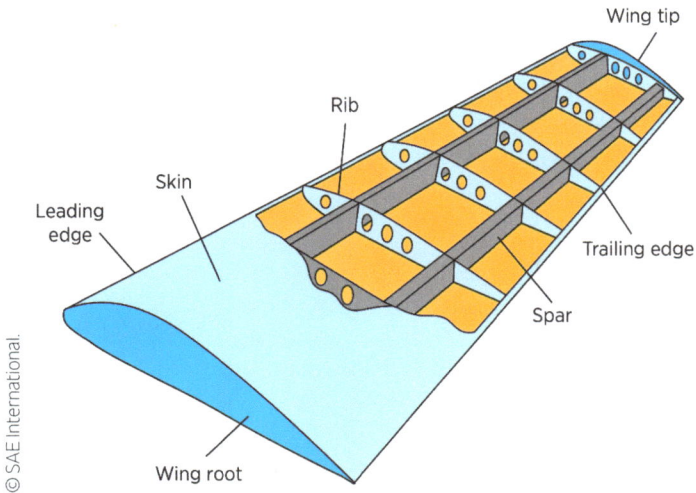

© SAE International.

3.2.14.
FOD

An FOD defect is when a small piece of composite material, either carbon fiber "fuzz-ball" or "resin ball" that has collected on surfaces of the head or other debris from the production area, falls onto the part during layup. This results in a small excess volume of material on the ply if laid up over (**Figures 3.21**, **3.22** and **Table 3.14**).

Figure 3.21 Rendering of a FOD model.

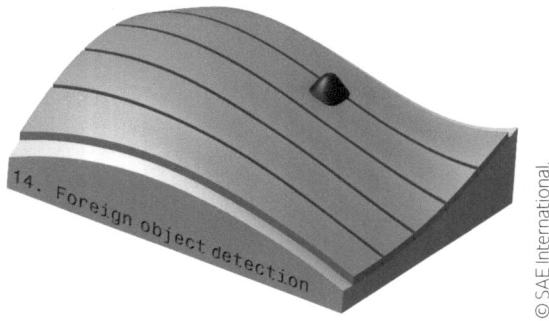

14. Foreign object detection

© SAE International.

Figure 3.22 Image of an actual FOD defect.

Table 3.14 FOD characteristics.

Anticipation	Monitoring the head of the machine and the production area for FOD and routine cleaning of surfaces are the appropriate ways to anticipate this defect.
Existence	Visually detect as it will be an irregular shape out of thickness along the same layer.
Significance	An FOD defect in layup can cause the portion of the next ply above the defect to improperly adhere to the defect's ply. This will lead to an undesired shape and strength of the part being made.
Progression	Unknowable as it will depend on the "foreign object."

© SAE International.

3.3.
3D Printing Defect Models

Knowledge and detection of defects require some expertise in the size, shape, and significance of the considerable number of possible defects. Detection then becomes increasingly difficult due to the substrate and incoming material both being a dark black color, when considering carbon fiber tows. The inability to detect defects leads to a part with poor quality. This section aims to provide an educational opportunity on these defects using 3D modeling and 3D printing to visualize each defect. To ease visualization, the model can be printed in three different colors as follows: First, the tool surface is printed in one color, and then, the tows are printed in a second color. The contrast of these two colors can then be used to visualize defects such as boundary coverage or tow drops. Defects associated with single tows such as gaps, overlaps, and twists can be modeled and printed using a third color. This method creates 3D models with easily identifiable defects that can be used to educate or train AFP personnel.

3.3.1.
Modeling

Modeling of the defects was done in multiple steps. First, the tool was created for the tows to be placed onto. After the tool was created, a set of five tows was modeled with each one being a separate body. The tows can be modeled with or without defects to demonstrate good- and bad-quality layups. Each of the defects is modeled in its own file and then combined into a single model for printing.

To begin modeling, a tool was created with a length and width of 100 mm and 75 mm, respectively, and was termed "base model" (**Figure 3.23**). The name of each model is embossed on the side of the tool to quickly know which model is in hand. A tool refers to the surface that the AFP head will place the tows onto. The geometry of the tool surface was created to demonstrate what would typically be seen on a complex tool. Complex tools have curvatures of varying values at each point along the tool. Such a geometry makes fiber placement difficult due to the defects it induces. The tool surface is then projected straight down to create the base of the tool where it is mounted to a flat surface or mandrel.

- Download link of 3D printed parts can be accessed with this quick-response (QR) code

Figure 3.23 3D model of the tool surface.

© SAE International.

Apart from the defect modeling, tool surface modeling can be used to teach personnel about the effects of geometry such as concavity or possible collision areas. Concave portions of a tool are prone to bridging defects, and severe curvatures are probable collision areas. Understanding the limits on tool geometry is the first step in producing high-quality AFP structures.

3.3.2.
Tow Modeling

As previously mentioned, tows are the strips of composite material that are placed onto the tool surface using the AFP machine and attached head. The tows are modeled following the geometry of the tool surface modeled in the previous step. Five large tows are used instead of smaller ones to ease visualization of defect and non-defect tows. In a real manufacturing environment, between 8 and 32 tows are usually deposited at once. The tows "blend" together, and it is challenging to decipher the appearance of minor defects. Learning what defects to look for and how to spot them on the larger models makes for an easier transition into locating defects in a real manufacturing environment.

Defect-free tows are created with equally sized small spaces between them to identify one tow from another, and these gaps are not to be interpreted as defects. To create each tow, straight lines are projected onto the tool surface using reference points on the edge of the tool to define the beginning and end. Each tow is a closed curve and, therefore, can be extruded to create the thickness of the tow. The tows can be created in the same file as the base, or they can be modeled in a separate file and combined with the base in a final model (**Figure 3.24**). The distinction between these two methods will become clear in Section 3.3 when the printing of the models is discussed.

Figure 3.24 Explanation of combining the base model with the non-defect tows.

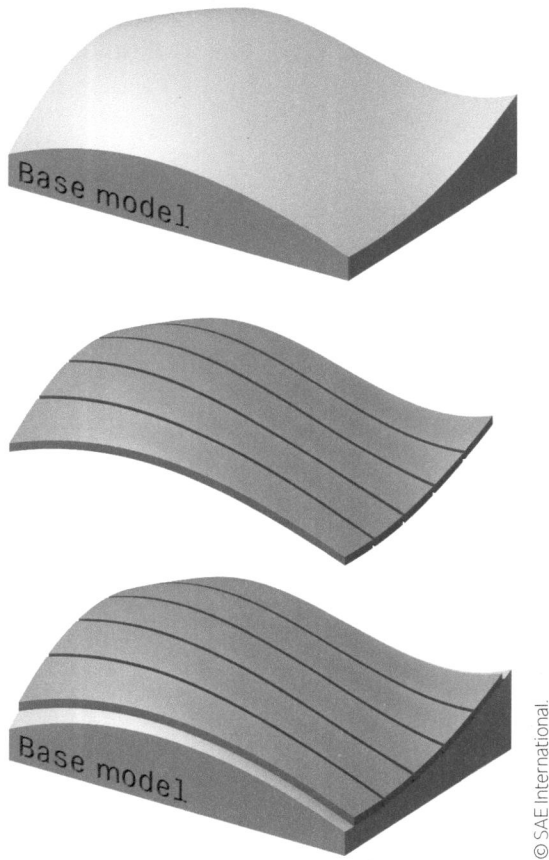

© SAE International.

Modeling tows with defects is a similar process to modeling tows without defects. The first step is deciding which tow or tows are going to have the specific defect. The defect is then modeled using the geometry of the previously created tool surface and surrounding tows. This process is demonstrated in **Figure 3.25** where a gap/overlap is modeled. In this model, the center tow is modeled to have a gap on one side and a resulting overlap onto the tow on the opposing side. Combining the defect and the other tows with the base model creates the complete gap/overlap model. The benefit of creating the tool and each tow separately will be clear when the 3D printing methods are discussed. Using the same method, any desired defect can be modeled.

Figure 3.25 Example of modeling (a) defect-free tows with a (b) gap/overlap.

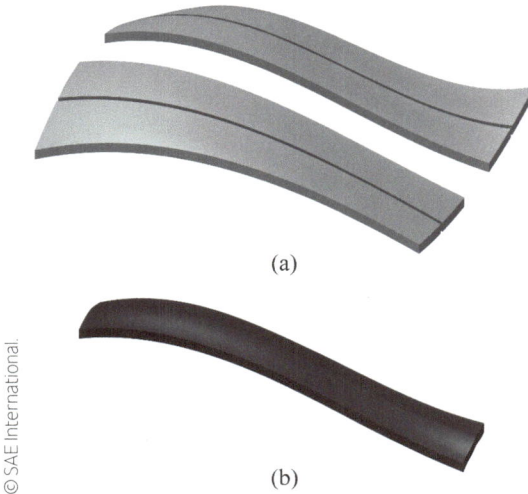

(a)

(b)

© SAE International.

Nomenclature

Duration: 22 minutes

Description: This video, part of "Composites A-Z: 30 Days of Composites," focuses on composite nomenclature, defining terms like matrix, fiber, lamina, and laminate. It covers fiber angles, stacking sequences, and laminate types (symmetric, balanced, cross-ply), while introducing tooling concepts (OML, IML) essential to understanding composites.

Scan the QR code below to watch the video.

3.3.3.
3D Printing

3D printing of these educational models creates a hands-on experience for personnel. Printing can be accomplished with any available printer and in two methods: (1) single print or (2) multiple prints. The single print method combines the tool and the tows into a single model for printing. Although this method is faster and easier, it results in the entire model being one color (unless using a multicolor printer). Using method 2, the model can be printed in multiple colors with the base, non-defect tows, and defected tows each being a separate color. Printing with multiple colors allows for the defects to be highlighted for easy detection. The method chosen is solely based on the desired appearance.

First is the single print method. In this method, the tows and tool will be combined into a single stereolithography (STL) file for printing. Depending on how the model was created, this may require creation of an assembly before exporting the model. Once exported, the STL model is imported into the desired slicer software that will create the individual layers and G-code for the printer to follow. The authors used the UltiMaker Cura slicer and have provided an example of two sliced models in **Figures 3.26** and **3.27**. Specific printer configuration properties such as nozzle diameter, layer height, and infill are not required. The chosen properties should be based on the knowledge of the printer's performance. If there is not a lot of experience with the printer, the default options will provide a quality print.

A major advantage of this method is that no support material is required, except for the defects that are not in contact with the tool surface. Support structures are not a part of the model and are generally created through a slicer software. These are generally used in places where the model is not supported by any underlying material; therefore, the printer has nothing to deposit the material onto. With the ability to use one continuous print with minimal support material, this method will be faster and more efficient.

Figure 3.26 Sliced models of a loose tow defect.

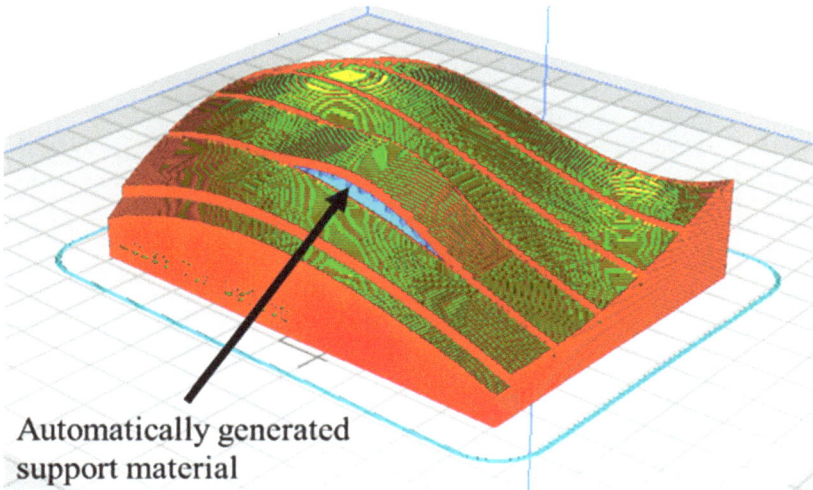

Automatically generated
support material

A. Brasington, T. Schachner and R. Harik, "3d Modeling and printing of automated fiber
placement defects," in CAMX – The Composites and Advanced Materials Expo. CAMX
Conference Proceedings, Orlando, Florida, 21-24 September 2020. Reprinted by permission
from the Society for the Advancement of Material and Process Engineering (SAMPE).

Figure 3.27 Sliced models of a boundary coverage defect.

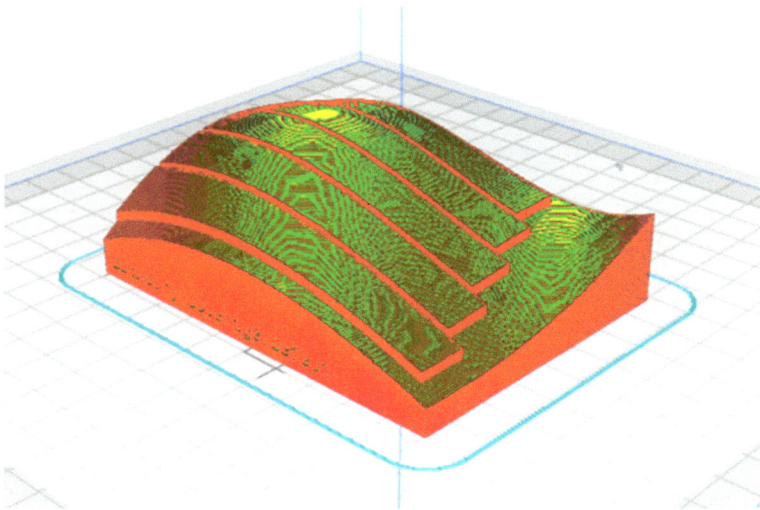

A. Brasington, T. Schachner and R. Harik, "3d Modeling and printing of automated fiber
placement defects," in CAMX – The Composites and Advanced Materials Expo. CAMX
Conference Proceedings, Orlando, Florida, 21-24 September 2020. Reprinted by permission
from the Society for the Advancement of Material and Process Engineering (SAMPE).

Observing the angle deviation defect, it is apparent that some defects are hard to notice in a model printed using a single color. To create a more visually appealing printed model, multiple prints, or a dual nozzle 3D printer, can be used with various colors to highlight the defect. However, this method will be more time- and labor-intensive due to the number of prints and assembly that are required. To begin, the base with the desired defect name embossed on the side should be printed. Once the base is printed, each tow associated with that defect should be printed separately. Depending on the tows being printed and the placement on the printer bed, support material will be required. If the printer's capabilities allow for multicolor printing, it may be possible to print all the tows in the two desired colors at one time.

Did You Know?

The mechanics of composites can be principally split into two categories: micro- and macro-mechanics. Together, these help us understand the mechanical response of a laminate under load. Understanding the properties of the constituents plays a vital role in comprehending micromechanics. This includes the incorporation of volume fractions, constituent properties, and fiber arrangement. However, macro-mechanics primarily deals with engineering aspects of composite material and its responses to applied loads. This level deals with the scale of laminates and incorporates the analyses accomplished through micromechanics. Analyses here define the behavior of a laminate such as stiffness and strength. A level up from macro-mechanics is a complete structural analysis where a laminate's properties are applied to an entire structure to predict its behavior under loads.

After the base and each tow are printed, they can be assembled to create the final model. The assembly is accomplished by using an adhesive to attach the tows to the tool. Each tow should fit precisely in place due to the matching curvature of the tool and the tow. Once assembled, the model should present a series of tows with a highlighted defected tow as it was modeled.

3.3.4.
Completed Defect Models

Following the modeling, printing, and assembly procedures outlined above, each defect previously detailed was modeled and printed using three different colors. An example of printed models can be seen in **Figure 3.28**.

Figure 3.28 Examples of some 3D printed defect models.

© SAE International.

3.4.
Conclusion

This chapter has described a way to categorize AFP defects and does so with a number of the major defects that have been identified through the author's experience and through a review of the available literature on AFP defects. In categorizing the defects in this way, the defects can be defined from the viewpoint of the designer, the process planner, the machine operator, and even the inspector. This allows for a rich understanding of each individual defect, which can help to fully answer the following questions:

- Can the formation of a certain defect be anticipated based on knowledge of the part geometry, machine parameters, and process planner decisions?
- Can the existence of a certain defect be confirmed based on available inspection systems such as profilometry, thermography, ultrasonic, or other technologies?
- Can the significance of the existence of the defect be understood, in a certain size/shape, on the overall integrity of my structure?
- Can defect progression be explained to the point of the performance effect and how it can initiate failure?

The presented methods for 3D modeling and printing of AFP defects can play an integral role in spreading the knowledge associated with these defects. This knowledge is typically gained through experience with inspecting numerous AFP-manufactured plies. Exposing personnel to these models introduces the types of AFP defects seen during manufacturing along with their geometry without having any prior experience. Identification of defects is just a small step in a complete understanding of the AFP process. Combining the education gained from modeling and printing the defects with their actual effects and significance will bring a broad knowledge of the process of inspecting and reworking tows.

References

3.1. Harik, R., Gurdal, Z., Saidy, C., Williams, S.J. et al., "Automated Fiber Placement Defect Identity Cards: Cause, Anticipation, Existence, Significance, and Progression," in *SAMPE 2018*, Long Beach, 2018, accessed March 24, 2020, https://www.researchgate.net/publication/326464139.

3.2. Brasington, A., Schachner, T., and Harik, R., "3D Modeling and Printing of Automated Fiber Placement Defects," in *Composites and Advanced Materials Expo, CAMX 2020*, Virtual, 2020.

3.3. Wehbe, R., Tatting, B., Harik, R., Gurdal, Z. et al., "Tow-Path Based Modeling of Wrinkling during the Automated Fiber Placement Process," in *The Composites and Advanced Materials Expo CAMX2017*, Orlando, 2017.

ADD Composites AFP-XS head at TU Delft's field lab SAM XL located in the Netherlands.

Chapter 04

Process Planning

Successful manufacturing is a direct result of good operational practices derived from detailed and extensive process planning. This planning stage should succeed, if not integrated into, the design phase. At this juncture, the matchmaking between design and manufacturing, including geometries, materials, and resources, takes place (see **Figure 4.1**). Process planning is defined as the creation of a manufacturing plan based on the working material, composite design, and manufacturing resources. It is one of the most critical and user-interactive portions of the AFP process. A process planner uses several tools, coupled with their extensive knowledge, to prepare the design for layup by the machine. This chapter compiles the findings from Harik et al. [4.1] and Halbritter et al. [4.2] and introduces the following steps to this initial work, with respect to laminate-level optimization (LLO).

The economic impact of process planning for AFP includes the time spent programming the machine before manufacturing, as well as labor and material costs that may arise during manufacturing due to poor or inefficient programming. Therefore, an improvement of this stage would allow for a more cost-effective and faster manufacturing pace to be set for the industry. To improve AFP process planning, we must first identify and discretize the functions that make up the process planning stage. By defining these individual functions, their total contribution to the result can be interpreted and steps can be taken to automate or semi-automate the most important and time-consuming ones. This would be a milestone to reduce the product development cycle time for advanced composite parts.

Figure 4.1 Process planning.

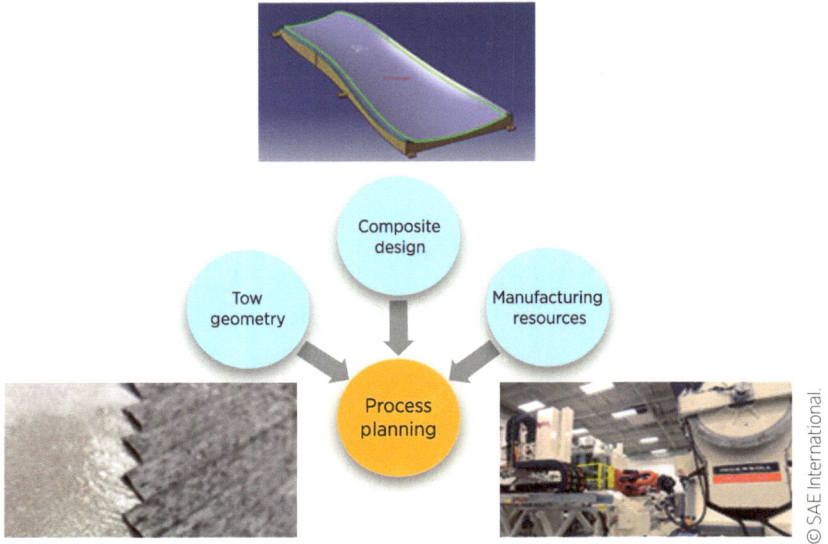

© SAE International.

The process planning functions in this chapter were specifically chosen from a down-selection process that took place after collecting input from process planners with industry experience as detailed in Harik et al. [4.1]. These individuals helped to guide the selection through the challenges they face and the decisions they make during the process planning portion of AFP, which consumes the most time and has the largest effect on the laminate. The functions that were the most difficult and time-consuming were then selected as the candidates to further investigate. Overall, the input given by these industry expert process planners helped to form a roadmap of functions that any process planner will have to face at some point during machine programming, and helped form an idea of the challenges we need to tackle to automate the most critical of these functions.

4.1.
AFP Manufacturing Plan

The creation of an AFP manufacturing plan is complex and requires intricate decision-making and engineering compromise. The creation of such a plan involves several steps that will be outlined throughout this chapter.

The fundamental principle of process planning is to establish a manufacturing configuration that most effectively produces a part closely matching the as-designed structure, while considering the broader goals of efficient, feasible, and inspectable production. This involves not only replicating the design as accurately as possible but also ensuring that the manufacturing process is fast, is achievable within existing capabilities, and results in a part that can be effectively inspected for quality and conformance to specifications. Before delving into each process planning function in detail, a flowchart is provided in **Figure 4.2** to summarize the process. This principle provides a comprehensive framework that guides the entire journey of a product from its initial design conception to its realization on the manufacturing floor. It not only outlines the sequential steps involved in this process but also emphasizes the importance of integrating design considerations with practical manufacturing constraints. By following this framework, manufacturers can ensure a seamless transition from design to production, minimizing costly errors and delays while maximizing efficiency and quality. The steps are broken down into design assessment, process planning, tool form manufacturing, dry run, and AFP manufacturing. This workflow has been tested in manufacturing and demonstrated its validity as a generic AFP manufacturing approach.

Figure 4.2 Modified process planning flowchart based on the work of Harik et al. [4.3].

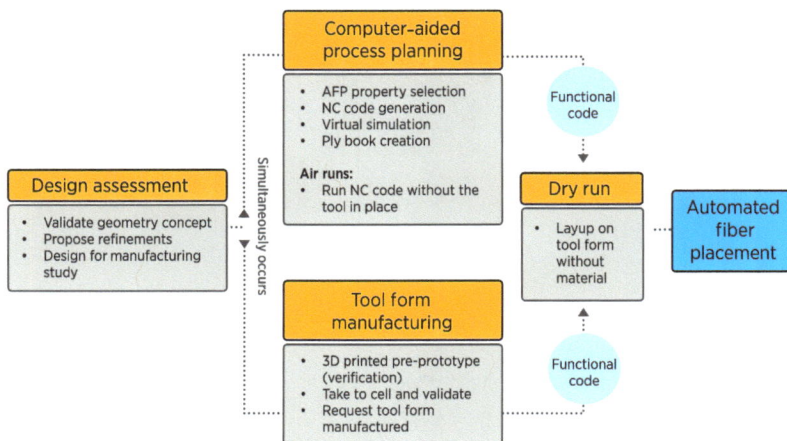

R. Harik, A. Lovejoy, D. Jegley and C. Yokan, "Thin-ply: Exploration and Manufacturing with Automated Fiber Placement," in CAMX Conference & Exhibition, Orlando, Florida, 21-24 September 2020. Reprinted by permission from the Society for the Advancement of Material and Process Engineering (SAMPE).

4.2.
Process Planning Functions

In this section, we list the different process planning functions that are executed by the process planner. Although presented sequentially, process planning involves a complex and iterative balancing act, where planners often alternate between and revisit functions as they strive to construct the optimal plan. Manually juggling these interconnected and sometimes conflicting objectives is a challenging endeavor, underscoring the need for computer-aided process planning (CAPP) with its multiobjective optimization capabilities to streamline and enhance this process.

AFP Process Planning

Duration: 35 minutes

Description: This video, part of "Composites A-Z: 30 Days of Composites," delves into AFP process planning. It covers layup strategies, reference curves (fixed angle, constant curvature, geodesic, variable angle), coverage strategies (offset, shifted, independent curves), and the role of computer-aided process planning (CAPP) in optimizing manufacturing quality.

Scan the QR code below to watch the video.

4.2.1.
Boundary Creation

The boundary is the general border where the layup must begin and end. It establishes a clear boundary between the desired extent of the layup, as determined by the process planner, and the restricted zone where machine collisions with obstacles on the mandrel could occur such as holes, bumps, or vacuum bag openings. The boundaries are typically defined through the design process; however, other boundaries can occur due to obstacles around the AFP machine. An example boundary is shown in **Figure 4.3**. Note that the external boundary of a part is typically larger than the final size of said part. This allows for excess material of lower quality to be cut from the part. Previous chapters have discussed the causes of lower quality at the beginning and end of courses.

Figure 4.3 Example of a boundary.

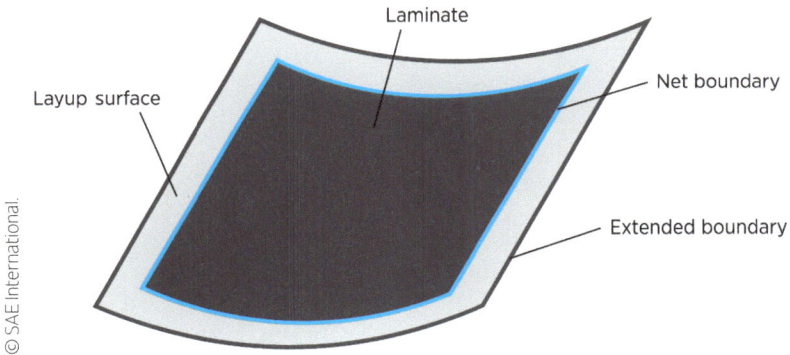

When considering the creation of ply boundaries, it is also critical to consider ply drops. When designing composite structures, often times a series of plies can have the same ply boundary. However, for manufacturing, such a stack of plies presents a layup issue because there is a large thickness change where the AFP machine cannot adequately place the tows. Neglecting this aspect can create a significant resin-rich zone devoid of fibers. This excess resin can weaken the composite, leading to delamination and ultimately structural failure. In certain material systems, abrupt changes in thickness can cause manufacturing challenges for AFP machines. To address this, gradual ply transitions, or "ramps," are sometimes incorporated to ensure consistent and controlled thickness buildup during the layup process (see **Figure 4.4**).

Figure 4.4 Diagram showing ply drops.

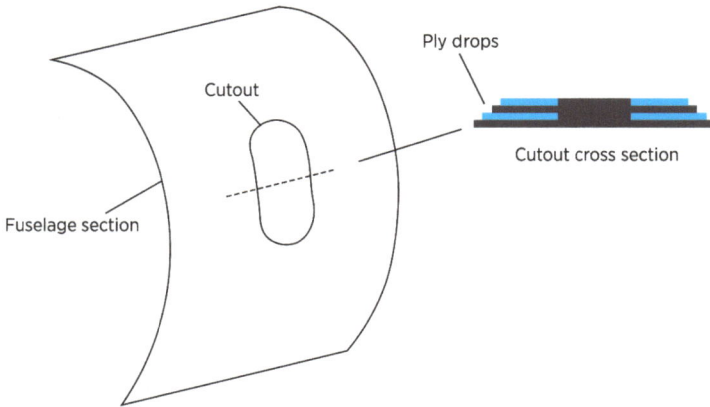

4.2.2.
Layup Strategy

When placing material over a complex surface, the fiber angles of the tow can vary because of the surface's geometry. Due to this effect, various layup strategies are employed to properly create tool paths. An essential part of defining a path is the creation of a reference curve that defines the initial path that will influence the remaining coverage of the part. The two main reference curve types that will be focused on in this chapter are fixed angle (also referred to as rosette and constant angle) and geodesic (also referred to as natural and constant curvature). The selection of these strategies is dependent on the desired outcome of the final part as well as an understanding of how the surface will affect the generation of the reference curve. Chapter 5 of this book will detail the different types of research strategies.

Did You Know?

Founded in 2011, the Ronald E. McNAIR Center for Aerospace Innovation and Research has a mission to grow South Carolina's knowledge-based economy and support industry through education, research, and industry advancement. All the research studies culminating in this book are conducted at the McNAIR Center. This chapter includes one of the principal research themes at the center: automation of process planning for composites initiated by Professor Ramy Harik. For more details about this topic, the readers are invited to read the dissertations of Joshua Halbritter and Alex Brasington and the thesis of Noah Swingle.

A fixed-angle strategy maintains a constant fiber angle along the surface. This type of strategy is often employed in industry due to the complete control of the fiber angles. Contrarily, a geodesic strategy follows the curvature of the surface regardless of deviation from the nominal fiber angle. Such a strategy is often employed to avoid steering since the curvature along a geodesic path is null. When dealing with relatively low curvature surfaces, the path using a geodesic strategy will not vary significantly from the intended fiber angle. The defined reference curve can then be propagated to cover the remaining surface. The typical coverage strategies include independent curves, parallel methods, and shifted methods. The independent curve method uses independently drawn reference curves to cover the surface. The parallel method propagates the curve by computing parallel curves that are equidistant from the previous one, and the shifted method simply shifts the curve by applying a perpendicular translation. A schematic visualizing the results of using these strategies is shown in **Figure 4.5**. The subsequent chapter will detail the mathematical formulation of various layup strategies.

Figure 4.5 Schematic of the effects of layup strategy selection.

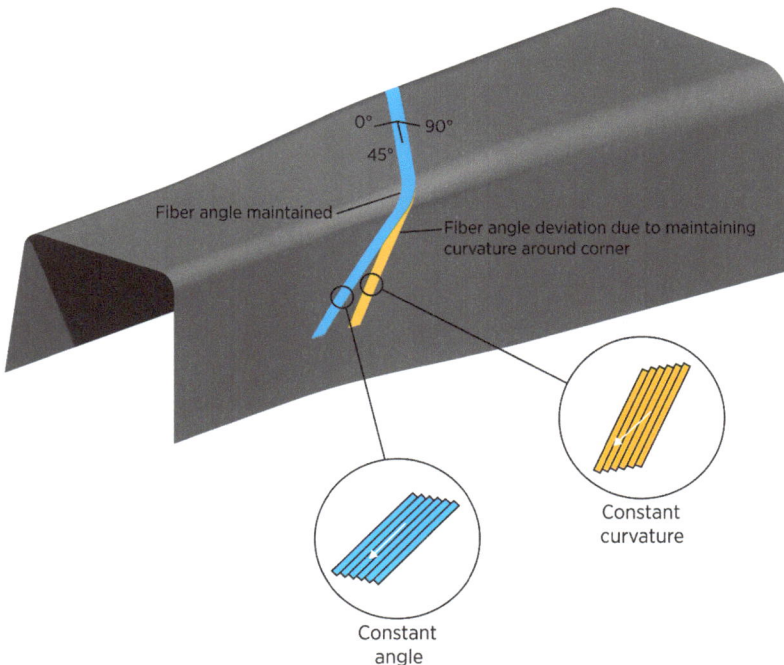

4.2.3.
Starting Point

The starting point indicates where the computational material coverage strategy will initiate from. The starting point can change depending on the complexity of the mandrel, the complexity of the part's geometry, and how much roller coverage there will be at a certain point. The starting point has an influence on the computation of the center path for the machine. Depending on the placement of the starting point, the resulting reference curve and propagated curves can have large variations. When dealing with a predefined reference curve, or guide curve, the starting point defines the location of the first course. **Figure 4.5** demonstrates the selection of a starting point.

The selection of starting points has no defined methodology. The determination of the starting point's location becomes increasingly difficult when considering a large structure with an excessive number of options. This is typically a manual trial-and-error process in which a process planner chooses a starting point and manually evaluates the performance of the resulting tow placement across the part. Even though this is a monotonous process, careful considerations must be taken when choosing starting points to ensure that unnecessary defects are not produced, defects are not located in structurally critical areas of the structures, and the locations of defects do not coalesce between subsequent plies (also known as defect staggering). The primary constraint for choosing a starting point is that it ideally lies within the current ply's boundary, although this requirement may vary depending on the specific programming algorithm or software used.

4.2.4.
Tow Width Definition

Defining tow width determines the width of material that will be used during manufacturing. In AFP, tows of 6.35 mm (1/4 in.) are most common. Smaller widths are desired for more complex tool geometries because they better conform and cover the tool, while less complex geometries allow for larger widths. AFP heads are often desired for a particular tow size limiting the variability allowed in this input. The tow width definition can also be modified within a process planning software to account for any error between the manufacturer's tow width specification and the actual width of the tow. The error in the tow width is usually found by the process planner if they verify the dimensions of the tow and find a difference between the specified width and the actual width. The final tow width is a critical parameter as it determines the width of each course laid down by the AFP machine. This course width, in turn, is influenced by the number of tows placed side by side, the spacing or gaps between these tows, and the specific construction and capabilities of the AFP machine itself. These factors collectively dictate the overall width of the composite material being produced and must be carefully considered. Larger courses result in lower manufacturing cycle times but may cause more defects due to less conformability.

4.2.5.
Steering Constraints

Steering constraints (and violations) are the restrictions a process planner will set on any steering necessary during the layup process. The main constraint is steering radius and will vary largely, depending on the path algorithm chosen for the part being fabricated and the curvature of the tool being laid up on, in addition to the composite's width and material properties. Tows used in AFP typically allow for more steering (lower steering radii) than wider tapes. Certain layup strategies can alleviate steering constraints but may result in other defects such as angle deviation, gaps, and overlaps. **Figure 4.6** shows the steering radii that the constraints are applied to. Note that this shows the steering radius measured at the course centerline, but it can also be measured at the tow centerlines for a higher fidelity measurement.

Figure 4.6 Methods to measure and analyze steering.

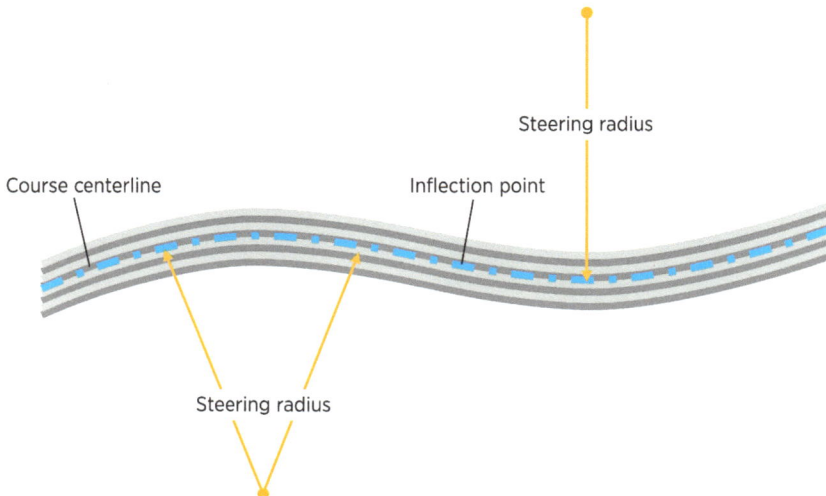

© SAE International.

4.2.6.
Interband Offset

An interband offset is an offset that is placed in between courses during a layup (see **Figure 4.7**). This offset will be used mainly to prevent any gaps or overlaps that may occur during layup and especially while steering, but it can also be used for any specific part geometry. The ability to input negative values in certain software allows for the creation of intentional overlaps using the interband feature, a result that might seem counterintuitive but might be needed for certain specific shapes/analyses. Many AFP systems can have overlaps between courses if no offset is applied. This is attributed to any minor imperfection in the material width tolerance or deviation from planned paths. Also, when placing material, the course can often become slightly wider than was assumed during planning due to material deformations. The offset alleviates these possible issues and ensures uniform thickness between parallel courses.

Figure 4.7 Interband offset (typically a positive value).

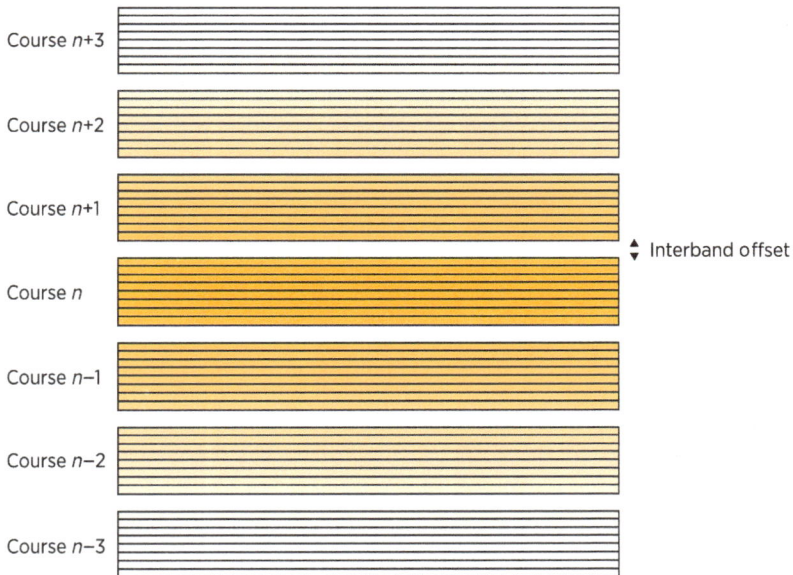

© SAE International.

4.2.7.
Stagger Shift

A stagger shift is when a process planner staggers two similarly oriented plies by shifting the top ply with respect to the bottom ply (see **Figure 4.8**). A stagger shift is usually used when gaps or overlaps are present in a ply so that these defects do not propagate to a subsequent ply oriented in the same direction. The shift is applied to the starting point and is moved perpendicular to the course travel direction. The stagger distance is typically a fraction of the course width with a good rule of thumb being one or two tow widths. Once the stagger achieves a complete course width, the defects are likely to move back to their original locations. This is due to the starting point being roughly located at the subsequent course centerline of the initial set of toolpaths before staggering. The application of stagger shifts is critical since gaps and overlaps that stack up between subsequent plies have a larger effect on structural performance than those that are randomly dispersed.

Figure 4.8 Stagger shifts.

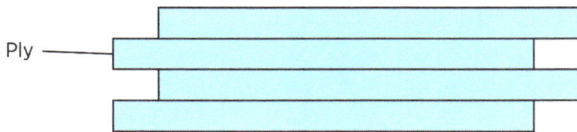

© SAE International.

4.2.8.
Boundary Coverage

Boundary coverage is when the material cannot perfectly meet up with the edge of a part when laying up at angles that are not normal to the boundary, which results in a gap or an overlap on that edge. The three common scenarios a process planner chooses from are a 100% gap–0% overlap, a 50% gap–50% overlap, or a 0% gap–100% overlap (see **Figure 4.9**). This is important, especially for internal boundaries as well (ply drops). The selection of a boundary coverage strategy is dependent on part specifications and requirements. The types of boundary coverage are seen often in variable stiffness designs where tows are added and cut within a part. This increases the importance of appropriate boundary coverage since the selection can increase or decrease the occurrence of gaps and overlaps.

Figure 4.9 Boundary coverage (0, 50, and 100%).

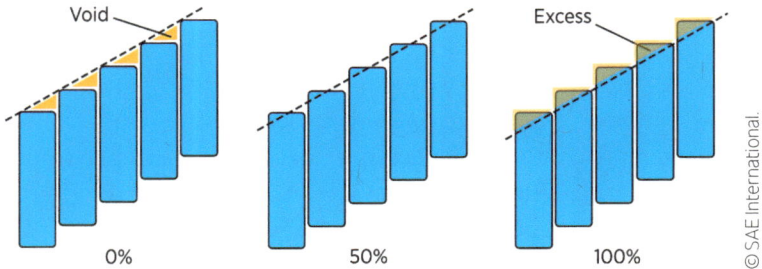

© SAE International.

4.2.9.
Dog Ear Addition

A "dog ear," also referred to as a "bird's beak" or "flag" depending on the shape of the excess material, is additional material laid up at any corner of a part where it is impossible to lay up perpendicular or parallel to that corner, as shown in **Figure 4.10**. Every machine has an absolute minimum tow length that it can layup, which causes dog ears to be present. The minimum tow length is typically set as the distance between the cutters and the nip line. This allows for material to be added out of the AFP head and placed before cutting. Without this minimum length, the tows may not leave the AFP head and can have significant wandering. The process planner will choose whether to implement these dog ears in the layup. If the ply boundary is external, the excess material will be trimmed off in postprocessing.

Figure 4.10 (a) Dog ear and (b) bird's beak addition.

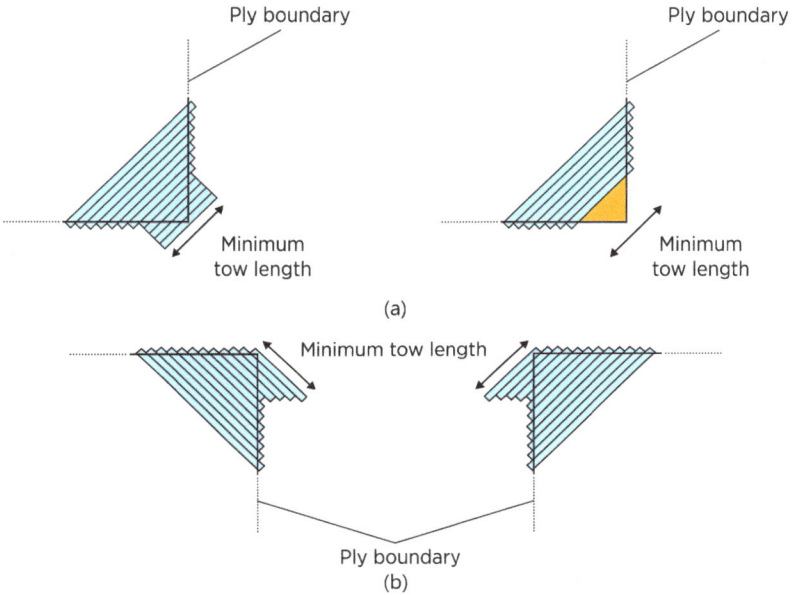

(a)

(b)

© SAE International.

Did You Know?

Process Planning in Subtractive Manufacturing

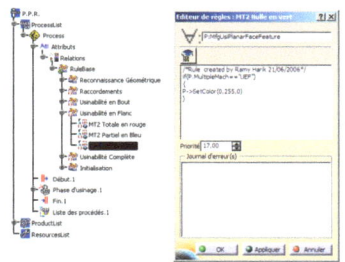

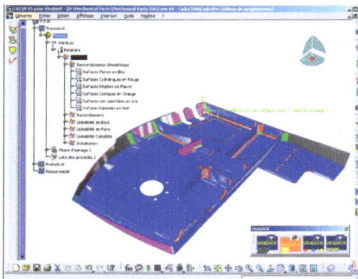

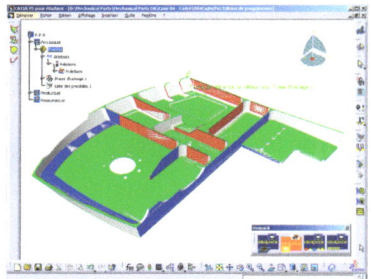

© SAE International.

The figure above shows the results from the USIQUICK research project, aimed at the automation of process planning functions in subtractive manufacturing. The images show the software (CATIA Add-on) launch, knowledge-based rules' definition, geometrical function results, and manufacturability end-milling results. The software was tested by process planners from Dassault Aviation. The process planners noted the following remarks: The software dramatically reduces the analysis and comprehension time of the part from days to a few hours, the software gives a powerful tool in its geometrical recognition function, where the visual output of the latter module helps the generation of a manufacturing fixture, the five-axis manufacturing direction function enables process planners to identify the manufac-turing fixtures without seeing the part itself, and the manufacturability function helps the fast generation of manufacturing strategies! More information can be read in *Ramy F. Harik, William J. E. Derigent & Gabriel Ris (2008) Computer Aided Process Planning in Aircraft Manufacturing, Computer-Aided Design and Applications, 5:6, 953-962, DOI: 10.3722/cadaps.2008.953-962* [4.4].

4.2.10.
Uni-/Bidirectional Layup

Unidirectional or bidirectional layup is the difference between laying up a course in one or two directions. In unidirectional layup, the AFP head does not rotate significantly between each course. However, in bidirectional layup, the AFP head must rotate 180° before beginning the next course. Different parts will benefit more from one or the other depending on the part's geometry and the off-part motion due to the complexity of the mandrel. For shorter courses, it is more beneficial to use a unidirectional layup because the time to stop the machine, rotate the head, and then begin the next course takes longer than returning to the start of the next course. A specific rule is not applicable here due to the variation in AFP systems. Optimal layup configurations should be found by simulating the toolpaths and determining the best combi-nation of machine motion and layup direction for the most efficient process.

Filament Winding

Duration: 22 minutes

Description: This video, part of "Composites A-Z: 30 Days of Composites," explores filament winding. It's an automated, open molding process for axially symmetric parts like helicopter blades, using prepreg or dry fibers. Fibers are resin-coated, mandrel-placed, and tension-controlled, enabling mid-volume production with lower labor costs.

Scan the QR code below to watch the video.

4.2.11.
Off-Part Motion

Off-part motion is the amount of movement the machine must make between laying up different courses while not in contact with the tool. The most efficient off-part motion will vary from machine to machine and from part to part. It is always desirable to limit off-part motion because this is the time that the machine is not placing any material. There are many settings that can be adjusted during process planning that define the off-part motion. For instance, approach and retract motions can vary based on collision avoidance and suitable placement areas. Off-part motion is a typical area of optimization that can increase the efficiency of a manufacturing process without altering the on-part tool paths. Similar to the layup direction, simulations should be performed to determine optimal off-part motion planning. Optimal motion configurations are comprised of the best layup direction and off-part motion between courses.

4.2.12.
Machine Speeds

The machine speeds are the layup, add, and cut zone speeds for tow placement while laying up a part. These speeds determine how quickly or slowly the machine will layup and at what rate the machine will start layup and cut material during the layup process. This function will target the identification of optimal feed rates based on the selected layup strategy. There are often speed-up and slow-down zones at the beginning and end of courses to ensure the AFP machine has reached a steady-state speed before beginning material placement. Slower speeds at the start and stop of courses can also help with placement quality in these zones. It is also possible to change the feed rates across a part. For example, a highly contoured part will have areas that require slower movement than others. Feed rates can also be limited by the kinematics of the AFP machine. To move at a given speed, it is required that the axes of the machine can adequately traverse from one point to another in the given time interval. Achieving higher feed rates is currently an area of research that will enable higher throughput of AFP parts.

4.2.13.
Axes' Weights

The axes' weights are how the different machine axes are weighted at any point throughout a layup. Assigning a higher relative weight to an axis indicates that it will generate more of the machine motion during travel. In weighing the axes differently throughout the layup, a process planner can create a more efficient layup depending on the necessary layup strategy and the complexity of the tool. Properly weighing the machine axes during the numerical code (NC) generation stage is crucial. It ensures efficient motion and minimizes cycle time by accounting for the specific kinematics of the target machine.

4.2.14.
Angle of Tilt

The angle of tilt for the machine is what defines the tilt of the head of the machine during layup. The tilt of the machine's head will vary depending on the geometry of the part and the complexity of the mandrel being laid up on in order to maximize roller coverage. This function is based on machine configuration and investigates extreme opposed tilts. The degree of tilt is often dependent on collision avoidance. If the machine cannot place the material with the head oriented normally to the surface due to a collision, a tilt angle must be introduced. Most machines are capable of tilting forward and backward parallel and perpendicular to the direction of motion.

4.2.15.
Course Efficiency

The course efficiency of a part is how efficiently the courses are laid up on a part due to their dimensions, which are usually reviewed before or during simulation. Course efficiency investigates where partially empty courses can be eliminated by merging with other partially empty courses (**Figure 4.11**). However, merging courses must not typically interfere with the imposed course stagger, which intends to minimize coincident course edges and defect stacking through the laminate.

Figure 4.11 Course efficiency showing you can achieve the same coverage with fewer courses.

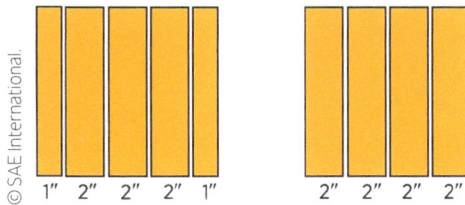

4.3.
Process Planning Automation

The authors have been deeply involved in process planning for AFP, specifically the automation of the process planning functions presented throughout this chapter. The involvement has resulted in a process planning automation tool deemed CAPP. The functions selected for automation were determined based on an industry survey in which process planners and engineers assessed their desire to automate each function and the antici-pated outcomes, as previously detailed [4.2]. From the survey discussed, five essential process planning functions for automation are extracted. These are layup strategy selection, starting point selection, stagger shift of starting point, dog ear creation, and steering constraints. The expectations and other remarks for each of the functions are provided in **Table 4.1**. These functions are the groundwork on which the CAPP software has been built.

Table 4.1 Essential process planning functions for optimization (reproduced from [4.2]).

Function	Expected output	Additional remarks
Layup strategies	• Provide a set of metrics for each layup strategy relevant to defect likelihood • Optimize based on user priorities such as angle deviation, steering, and gaps/laps	• Determined by engineering with some inputs from numerical control programming
Starting point	• Minimizes steering error, gaps, overlaps, and angle deviation, in addition to meeting course stagger requirements by placement of the starting point	• Finding proper placement varies by path generation algorithm and can be a tedious, but process critical action
Stagger shift	• Reduces coincident laps/gaps/seams through laminate thickness through modification of the starting point	• Considers the importance of reducing the feature that stagger shift is attempting to minimize • Combines stagger shift functionality with the start point function
Dog ears	• Modify ply boundaries to account for additional material placement • Determine optimal dog ear strategy by structural analysis	• Accept default dog ear method defined by process planning if they are not engineering-dependent
Steering constraints	• Dynamically vary the acceptable steering radius over the surface depending on local curvature • Recommend minimum steering radius based on geometry and layup orientation	• Only suggest minimum steering radius as guideline, since the surface curvature will alter the effects of in-plane curvature

J. Halbritter, R. Harik, C. Saidy, A. Noevere and B. Grimsley, "Automation of AFP Process Planning Functions: Importance and Ranking," in SAMPE 2019 Conference & Exhibition, Charlotte, North Carolina, US, 20 – 23 May 2019. Reprinted by permission from the Society for the Advancement of Material and Process Engineering (SAMPE).

The CAPP software was developed to improve and accelerate the process planning workflow for AFP. A previous release of the CAPP software is shown in **Figure 4.12**. CAPP assists process planners in identifying optimal starting point location and layup strategy for each ply of a laminate and is in its initial development stage.

Figure 4.12 Screenshot of the CAPP software.

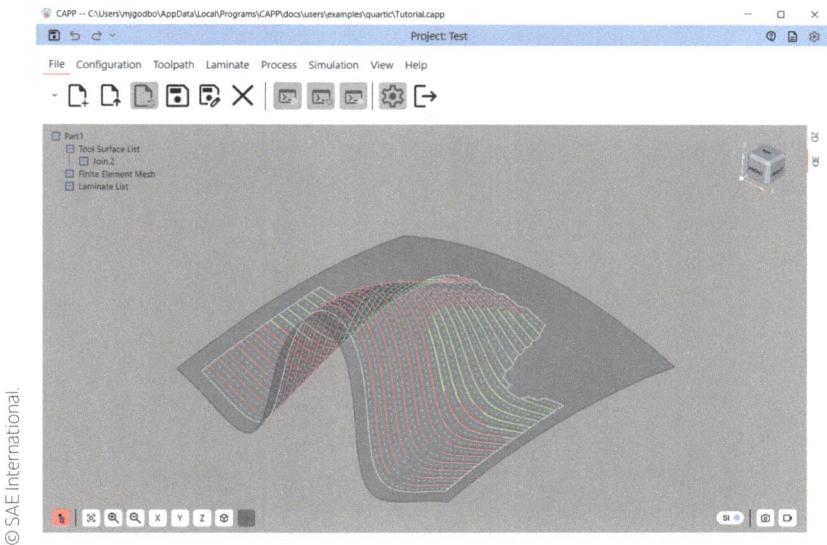

The process planning inputs are evaluated through communication with VERICUT Composite Programming (VCP) where necessary information such as starting points, ply boundaries, and layup strategies is passed to VCP. The course generation and defect analysis are performed by VCP and then given back to CAPP for further analysis.

Composites Design

Duration: 38 minutes

Description: This video, part of "Composites A-Z: 30 Days of Composites," explores the ABD matrix in composite analysis. It explains the 6x6 matrix linking loads to strains, combining A (in-plane), B (coupling), and D (bending) matrices, derived from stiffness and layer properties, highlighting coupling effects and stacking sequence impacts.

Scan the QR code below to watch the video.

4.3.1.
Defect Analysis

The CAPP software is designed to analyze the occurrence and distribution of key AFP defects that can be predicted based on the geometry of a layup. These defects include area-based defects: gaps and overlaps, along with point-based defects: angle deviation and steering. The analysis method provides the capability to incorporate the four defects into a single scoring metric. To begin, there is a threshold value that must be chosen by the process planner for each defect type. The threshold value represents the minimum threshold value for the defect, meaning the analysis will disregard defects smaller than this value. Utilizing the predicted defect data, instance and severity values can be calculated. The instance value is the number of defect instances out of the total defects that fall above the threshold value. Severity then measures the accumulated defect values above the threshold with respect to the total amount of each defect. In short, instances capture the number of defects, while severity attempts to account for the impact of the defects. The thresholds, instances, and severities are summarized with a table within the CAPP software resembling the one shown in **Table 4.2**. Choosing the threshold values takes some experience with process planning; however, the ones seen below are commonly used. Gap and overlap thresholds can be either an area or width value. The values in the instance and severity columns are scores (over 1).

Table 4.2 Example feature threshold table.

	Threshold	Instance	Severity
Gap	300 mm²	0.468	0.946
Overlap	300 mm²	0.209	0.514
Angle deviation	2°	0.369	0.860
Steering radius	2000 mm	0.459	0.336

As mentioned previously, defect types to be considered here can be either area defects or point defects. Therefore, two methods for assessing the instance and severity metrics are required. The bounding polygon of area defects, along with the respective attributes, is used for determining instances and severities of area defects. Using the threshold value for gaps and overlaps, the instance is computed by counting the number of polygons that exceed the threshold and dividing by the total polygons. The threshold can be assigned based on area, width, or length of the defect. The severity then accumulates all the specified attributes above the threshold and divides them with the total sum of the attributes for each defect. Recall that each point defect is described by its coordinates and a corresponding defect value. Utilizing the thresholds for the point defects, the instance is calculated by counting the number of points that exceed the threshold and dividing by the total number of points. The severity then accumulates the defect value at each point exceeding the threshold and is divided by the sum of all defect values for each defect.

A comparison and ranking process is used to provide a method for creating an overall ranking of many features through a series of pair-wise comparisons through the analytic hierarchy process (AHP) [4.5]. AHP works by allowing an engineer to decide how severe each defect is in comparison with other defects. By doing pair-wise comparisons between each defect, an overall ranking for how important each defect is can be created. Using the final overall defect rankings and the measured defect instances and severities, a score can be assigned to each ply.

Utilizing the weights and measured values, the final score can be calculated. Each of the weights is multiplied by the measured value to obtain a score for the respective feature. One minus the sum of these scores results in the overall score of the ply. A score of 1 is the optimal score and would mean that there are no defects present that exceed the provided thresholds. An example 2D scoring is shown in **Figure 4.13**.

Figure 4.13 Schematic of the process for identifying and measuring stacked defects.

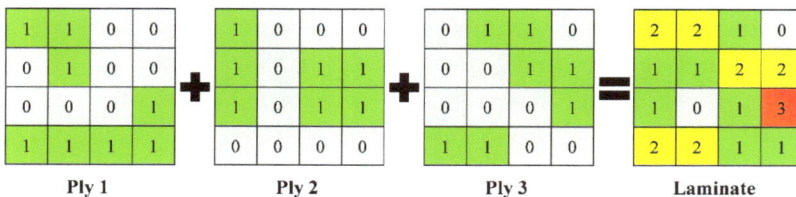

1	1	0	0
0	1	0	0
0	0	0	1
1	1	1	1

Ply 1

+

1	0	0	0
1	0	1	1
1	0	1	1
0	0	0	0

Ply 2

+

0	1	1	0
0	0	1	1
0	0	0	1
1	1	0	0

Ply 3

=

2	2	1	0
1	1	2	2
1	0	1	3
2	2	1	1

Laminate

Halbritter, J., Swingle, N., Wehbe, R., & Harik, R. (2022). Minimizing Through Thickness Defect Stack-Up for Automated Fiber Placement of Composite Laminates via Fiber Path Optimization. SAMPE 2022. Charlotte, NC. Reprinted by permission from the Society for the Advancement of Material and Process Engineering (SAMPE).

4.3.2.
Laminate-Level Analysis

Once individual plies and the resulting defects have been defined, they must be combined to form the full laminate. The nomenclature for a set of ply-level inputs will be a laminate scenario. During modeling and analysis on the ply level, course-to-course gap and overlap defects are extracted relative to their locations in 3D space. As plies are placed, these defects can begin to stack up on each other. These laminate-level defects have a larger effect on ultimate strength than on the ply level. To detect interactions between individual defects in two neighboring plies, even before the actual interaction is computed, each defect location must be checked with the nearness of every other defect location in the other ply. Therefore, naively for n plies having m defects each, m^n comparisons must be performed to even determine which ply defects will interact before the interaction is computed. The problem becomes even more complex with the defects represented in 3D, where tolerances with the actual representation of the defects complicate the interaction calculation. The number of comparisons may be reduced through different approaches as described in Halbritter et al. [4.6].

4.3.3.
Optimization Strategies

For ply-level optimization (PLO), the essential inputs to the optimization are the coordinates of the starting point along with the layup strategy. This set of inputs, combined with collisions and course efficiency, has shown itself to be the most influential in terms of ply manufacturability. The objective function for the PLO is the defect analysis methodology within CAPP. Recall that the function incorporates overlaps, gaps, angle deviation, and steering into a single score in the range [0, 1]. The function also has customizability through inputs into the AHP matrix that defines the importance of each defect's instance and severity. Such customization results in a modular algorithm that can be fine-tuned for any scenario.

Two methods for PLO are available with CAPP. First, optimization can be accomplished manually. This manual method allows a process planner to assess the results of the optimization and choose the best inputs for the given situation. The optimization is initialized with a grid of points with each point having a corresponding layup strategy. The user can select multiple layup strategies to evaluate the strategies at each location of the grid. The points can be expanded to search locally with an array of scenarios of a specified number and size. The array then moves in the direction of the point with the best score. Current work is trying to automate this process to behave in a gradient descent fashion. The iterations continue until a convergence, or the process planner deems the solution sufficient.

The second method for optimization is a semi-autonomous Bayesian optimization (BO) algorithm. Before outlining the optimization algorithm, it is critical to note that the objective function for the PLO is computationally expensive to calculate. Before a score can be computed, it is required that VCP compute all paths across a given ply. The time for this computation is highly dependent on available computational power, part size, and course width. All paths must also be computed for every iteration of the algorithm, greatly increasing the computation time. This computational weight will be considered through the utilization of a BO algorithm that incorporates a Gaussian process regression (GPR) machine learning model to reduce the required number of function calls to achieve optimality. Fine-tuning with manual iterations can then be performed as a last step to select the desired output and eliminate the need for complete convergence.

Once the PLO workflow is completed, users will have generated numerous different ply scenarios for each ply. Each ply scenario constitutes a specific starting point location and path geometry, which subsequently determine the geometrically induced defect distribution. The goal of LLO is to identify the optimal combination of ply scenarios that form a laminate with the least amount of through-thickness defect stacking. Specifically, LLO seeks to minimize the interactions of defects through the thickness of a laminate.

Application

The ABD matrix is a 6×6 matrix that serves as a connection between the applied loads and the associated strains in the laminate. Essentially, it defines the elastic properties of the entire laminate. This matrix is the cornerstone of any composite laminate analysis. It is composed of three separate 3×3 matrices: A, B, and D. Each of these can be calculated with the following equations:

$$A_{ij} = \sum_{k=1}^{N} \left(\bar{Q}_{ij}\right)_k \left(z_k - z_{k-1}\right)$$

$$B_{ij} = \frac{1}{2} \sum_{k=1}^{N} \left(\bar{Q}_{ij}\right)_k \left(z_k^2 - z_{k-1}^2\right)$$

$$D_{ij} = \frac{1}{3} \sum_{k=1}^{N} \left(\bar{Q}_{ij}\right)_k \left(z_k^3 - z_{k-1}^3\right)$$

Assembling these together and relating them to the applied forces and resulting strains and curvatures give

$$
\begin{bmatrix} N_x \\ N_y \\ N_{xy} \\ M_x \\ M_y \\ M_{xy} \end{bmatrix}
=
\begin{bmatrix}
A_{11} & A_{12} & A_{16} & B_{11} & B_{12} & B_{16} \\
A_{12} & A_{22} & A_{26} & B_{12} & B_{22} & B_{26} \\
A_{16} & A_{26} & A_{66} & B_{16} & B_{26} & B_{66} \\
B_{11} & B_{12} & B_{16} & D_{11} & D_{12} & D_{16} \\
B_{12} & B_{22} & B_{26} & D_{12} & D_{22} & D_{26} \\
B_{16} & B_{26} & B_{66} & D_{16} & D_{26} & D_{66}
\end{bmatrix}
\begin{bmatrix} \varepsilon_x^0 \\ \varepsilon_y^0 \\ \gamma_{xy}^0 \\ \kappa_x^0 \\ \kappa_y^0 \\ \kappa_{xy}^0 \end{bmatrix}
$$

Using these equations, we will find the [ABD] matrix of the following graphite/epoxy cross-ply laminate: [0/90]$_s$. The thickness of each ply is 5 mm with properties E_1 = 181 GPa, E_2 = 10.3 GPa, v_{12} = 0.28, and G_{12} = 7.17 GPa. We will also assume v_{12} = v_{21}.

$z_1 = -10mm$	
	0
$z_1 = -5mm$	
	90
$z_1 = 0mm$	
	90
$z_1 = 5mm$	
	0
$z_1 = 10mm$	

We can use the following equations for $\bar{Q}$ from previous examples to compute the matrix for the 0 and 90° plies:

$$\bar{Q}_0 = \begin{bmatrix} 1.82 * 10^{11} & 2.90 * 10^9 & 0 \\ 2.90 * 10^9 & 1.03 * 10^{10} & 0 \\ 0 & 0 & 7.17 * 10^9 \end{bmatrix} Pa$$

$$\bar{Q}_{90} = \begin{bmatrix} 1.03 * 10^{10} & 2.90 * 10^9 & 0 \\ 2.90 * 10^9 & 1.82 * 10^{11} & 0 \\ 0 & 0 & 7.17 * 10^9 \end{bmatrix} Pa$$

We can then compute the A, B, and D matrices as follows:

$$[A] = \bar{Q}_0 (z_2 - z_1) + \bar{Q}_{90} (z_3 - z_2) + \bar{Q}_{90} (z_4 - z_3) + \bar{Q}_0 (z_5 - z_4)$$

$$= \begin{bmatrix} 1.92 * 10^9 & 5.79 * 10^7 & 0 \\ 5.79 * 10^7 & 1.92 * 10^9 & 0 \\ 0 & 0 & 1.43 * 10^8 \end{bmatrix} N/m$$

$$[B] = \frac{1}{2} \left(\bar{Q}_0 (z_2^2 - z_1^2) + \bar{Q}_{90} (z_3^2 - z_2^2) + \bar{Q}_{90} (z_4^2 - z_3^2) + \bar{Q}_0 (z_5^2 - z_4^2) \right)$$

$$= \begin{bmatrix} 0 & 0 & 0 \\ 0 & 0 & 0 \\ 0 & 0 & 0 \end{bmatrix} N$$

$$[D] = \frac{1}{3} \left(\bar{Q}_0 (z_2^3 - z_1^3) + \bar{Q}_{90} (z_3^3 - z_2^3) + \bar{Q}_{90} (z_4^3 - z_3^3) + \bar{Q}_0 (z_5^3 - z_4^3) \right)$$

$$= \begin{bmatrix} 1.07 * 10^5 & 1.93 * 10^3 & 0 \\ 1.93 * 10^3 & 2.12 * 10^4 & 0 \\ 0 & 0 & 4.78 * 10^3 \end{bmatrix} N \cdot m$$

As a result,

$$[ABD] = \begin{bmatrix} 1.92 * 10^9 & 5.79 * 10^7 & 0 & 0 & 0 & 0 \\ 5.79 * 10^7 & 1.92 * 10^9 & 0 & 0 & 0 & 0 \\ 0 & 0 & 1.43 * 10^8 & 0 & 0 & 0 \\ 0 & 0 & 0 & 1.07 * 10^5 & 1.93 * 10^3 & 0 \\ 0 & 0 & 0 & 1.93 * 10^3 & 2.12 * 10^4 & 0 \\ 0 & 0 & 0 & 0 & 0 & 4.78 * 10^3 \end{bmatrix}$$

References

4.1.   Harik, R., Halbritter, J.A., Jegley, D., Grenoble, R. et al., "Automated Fiber Placement of Composite Wind Tunnel Blades: Process Planning and Manufacturing," in *International SAMPE Technical Conference*, Charlotte, May 2019, 1-16, https://doi.org/10.33599/nasampe/s.19.1538.

4.2.   Halbritter, J., Harik, R., Saidy, C., Noevere, A. et al., "Automation of AFP Process Planning Functions: Importance and Ranking," in *International SAMPE Technical Conference*, Charlotte, May 2019, https://doi.org/10.33599/nasampe/s.19.1592.

4.3.   Harik, R., Lovejoy, A., Yokan, C., and Jegley, D., "Thin-Ply: Exploration and Manufacturing with Automated Fiber Placement," in *CAMX 2020*, Virtual, 2020.

4.4.   Harik, R.F., Derigent, W.J.E., and Ris, G., "Computer Aided Process Planning in Aircraft Manufacturing," *Computer-Aided Design and Applications* 5, no. 6 (2008): 953-962, doi:10.3722/cadaps.2008.953-962.

4.5.   Saaty, R.W., "The Analytic Hierarchy Process—What It Is and How It Is Used," *Mathematical Modelling* 9, no. 3–5 (1987): 161-176.

4.6.   Halbritter, J., Swingle, N., Wehbe, R., and Harik, R., "Minimizing through Thickness Defect Stack-Up for Automated Fiber Placement of Composite Laminates via Fiber Path Optimization," in *SAMPE 2022*, Charlotte, 2022.

Ingersoll Machine Tools AFP machine at the McNAIR Center in Columbia, South Carolina.

Chapter 05

Layup Strategies

One of the principal tasks needed to effectively manufacture a composite structure with AFP requires knowledge-based decisions on the desired layup strategy. These decisions are typically inherent to different industries and usually stem from internal best practices. While some industries might set their goal to maximize coverage (minimizing gaps) or even coverage (minimizing both gaps and overlaps), others might accept reduced coverage as long as other defects (i.e., steering or angle deviation) are lowered below set thresholds. While these strategies are considered in the realm of manufacturing, they are vital to assess the design margins of safety. This is expanded on in Chapter 8 (Design for Manufacturing). Layup strategies alter the original design that was used to compute the part functional requirements. As such, it is expected to assess the influence of the layup strategy selection to ensure all design requirements are still met.

Figure 5.1 depicts the influence of process planning functions, particularly layup strategies, on the generation of a manufacturing ready design. Once an initial design meets the functional requirements, it still must have a validated layup strategy. The as-planned design is reassessed through the functional requirements to ensure validation. Only then can the design be considered as manufacturing ready.

Figure 5.1 Influence of process planning on design.

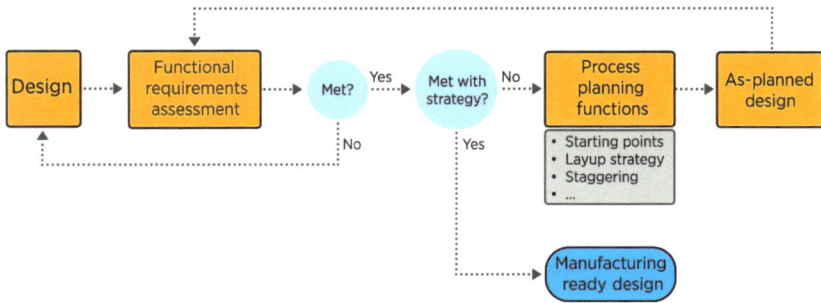

This chapter presents the different layup strategies that are existent in typical AFP toolpath generation software. It is adapted from the work of Rousseau et al. [5.1]. Layup strategies require the orchestration of (1) starting point selection, (2) reference curves, (3) coverage strategies, and (4) through-thickness staggering. This chapter is organized as follows: Section 5.1 introduces the fundamentals of starting point selection as well as research efforts to automated starting point selection, Section 5.2 details the computation of reference curves, Section 5.3 elaborates on coverage strategies that use the reference curve and propagate it through different techniques across the ply, Section 5.4 presents initial efforts on through-thickness staggering, and Section 5.5 includes both a summary and discussion on layup strategy generation.

5.1.
Starting Point Selection

When generating a reference curve with the strategies mentioned above, a starting point is needed to define where the reference curve should start (see **Figure 5.2**). The chosen layup strategy then propagates from this point to create the complete path. Depending on the placement of the starting point, the resulting reference curve and propagated curves can have large variations. These variations may lead to defects occurring within the laminate, resulting in differing properties when comparing the as-manufactured part with the as-designed part. When dealing with a predefined reference curve, or guide curve, the starting point defines the location of the first course. The guide curve is then interpolated to this location on the surface to generate the path.

Figure 5.2 Schematic of the effects of layup strategy selection.

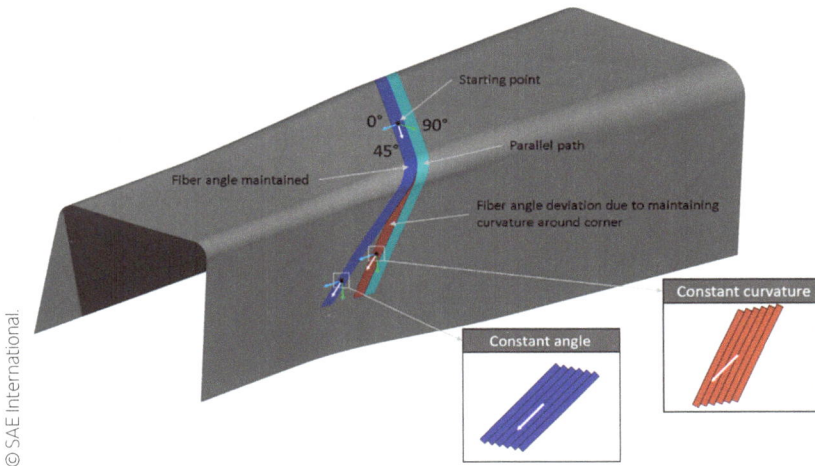

The selection of starting points has no defined methodology. The determination of the starting point's location becomes increasingly difficult when considering a large structure with an excessive number of options. This is typically a manual trial-and-error process in which a process planner chooses a starting point and manually evaluates the performance of the resulting tow placement across the part. Even though this is a monotonous process, careful considerations must be taken when choosing starting points to ensure that unnecessary defects are not produced, defects are not located in structurally critical areas, and the locations of defects do not coalesce between subsequent plies.

Did You Know?

Flank Milling in Subtractive Manufacturing

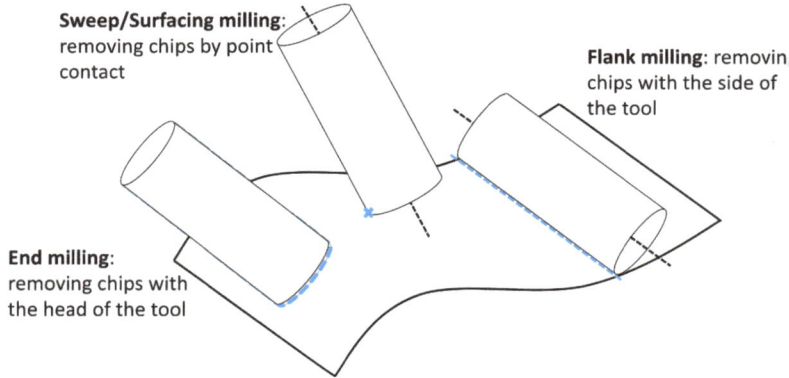

Sweep/Surfacing milling: removing chips by point contact

Flank milling: removin chips with the side of the tool

End milling: removing chips with the head of the tool

The figure above shows the different manufacturing strategies in subtractive manufacturing: end milling, flank milling, and sweep milling. The easiest strategy to compute is typically the end milling, as the chips are removed with the end of the tool. This also renders the process simpler to compute the toolpath trajectory as well as the final result, without worrying about potential tool deflection (to a certain extent). Sweep milling is typically removed by point contact, where manufacturing is conducted with the corner of the tool. Flank milling is a rather complicated process for multiple reasons. We will make a distinction of those reasons into two categories: toolpath generation and manufacturing. Flank milling is when you are using the side of the tool to execute the manufacturing process. As such, with respect to toolpath generation, planar surfaces can have an indefinite amount of accessibility directions. That can be computationally expensive to find the optimal set of orientations. Readers can refer to Harik's Thesis for details on potential algorithms. With respect to manufacturing, the length of the tool can deflect based on the different contact lengths. This needs to be accounted for, especially since the target of using flank milling is to actually generate better surface finishing, so accuracy must not be sacrificed.

5.2.
Reference Curves

The different courses laid up by the AFP process need to be placed in a way that the fiber orientations meet the required specifications. Most of the time, it is asked to make plies with fibers having well-defined angles 0, 90, 45, and −45° [5.2] as they provide a quasi-isotropic behavior for the structure. The 0° angle direction can follow the longest dimensions of the surface, as in **Figure 5.3**. It is simple to lay up plies with these angles on a flat panel, however, on complex surfaces, it is more difficult as the fiber angles in the tow can change due to the geometry of the surface. Also, it becomes difficult to define a fiber angle as a single rosette axis may not be suitable for an entire geometry. This is why many different layup strategies appeared, as one cannot use the same strategy to obtain the same structure as the tooling surface changes.

Figure 5.3 Flat panel with different fiber angles.

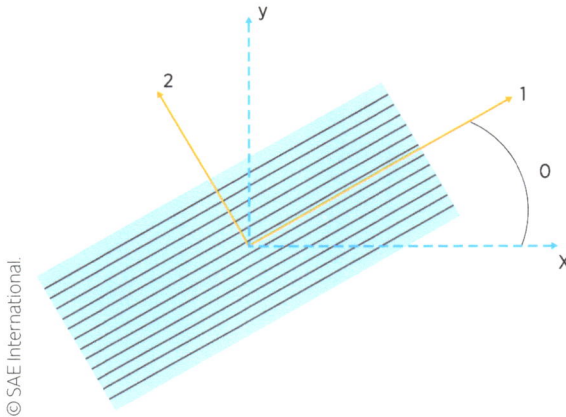

© SAE International.

Before covering the surface, an initial (or reference) curve is needed. In this section, the different strategies to find the reference curve will be detailed. Either parametrical approaches, or the use of a mesh, can be found in the literature. Both methods have advantages and drawbacks, which have been summed up by Chen, Fuhlbrigge, and Li [5.3] for automated spray painting. A mesh provides useful information, such as the different areas of the facets and their normal, which are important to generate the toolpath along the course. Nevertheless, a mesh is an approximation of the surface, so the precision obtained depends on how accurate the mesh is. One can keep in mind that a more refined mesh would increase the computation time. Using a parametrical approach, the surface would be known more precisely. In this section and the following ones, we will detail different strategies to find reference curves.

Another condition to check before finalizing the path design is the turning radius or the curvature of the path. Since the tows used in the AFP process have a finite width, the edges of the tow will be either under tension or compression while trying to adhere to a rectangular-shaped tow along a curved path. This mismatch in length between the tow and the actual path on the surface results in excess material buckling out of plane to form a wrinkle (**Figure 5.4**) on the compressive edge of the tow. As for the tensile side, the shortage of material will push the fibers to move closer to the center line leading to tow straightening, or in severe cases to move out of plane and fold over [5.4]. To avoid these defects, a minimum steering radius must be set beforehand depending on the material to be used. Usually, the minimum steering radius is determined experimentally by trying different radii of curvature with different combinations of process parameters (speed, temperature, and roller pressure).

Figure 5.4 Most common tow steering defects [5.5].

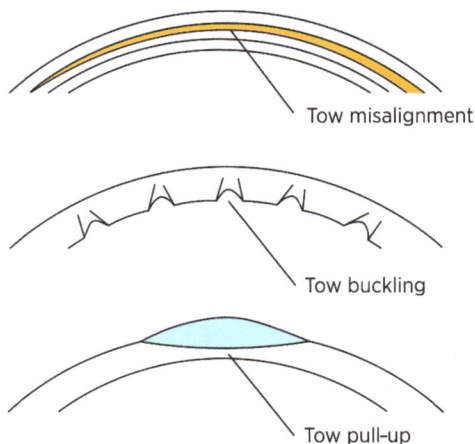

Tow misalignment

Tow buckling

Tow pull-up

Used with permission of Elsevier Science & Technology Journals, from Composites. Part B, Engineering, Lukaszewicz, Dirk H.-J.A.; Ward, Carwyn; Potter, Kevin D., 43, 3, 2012; permission conveyed through Copyright Clearance Center.

The stark contrast that can occur when selecting different layup strategies is highlighted in the cone case study shown in **Figure 5.5**. Each selection demonstrates significant variations in fiber angle and steering radius, which are key considerations when making such a selection. The generation and purpose of these reference curves will be detailed throughout the following subsections (**Figure 5.6**).

Figure 5.5 Constant curvature, geodesic, and constant angle paths on a cone with fiber orientation of the small and big radius equal to 45°: paths on the 3D cone [5.6].

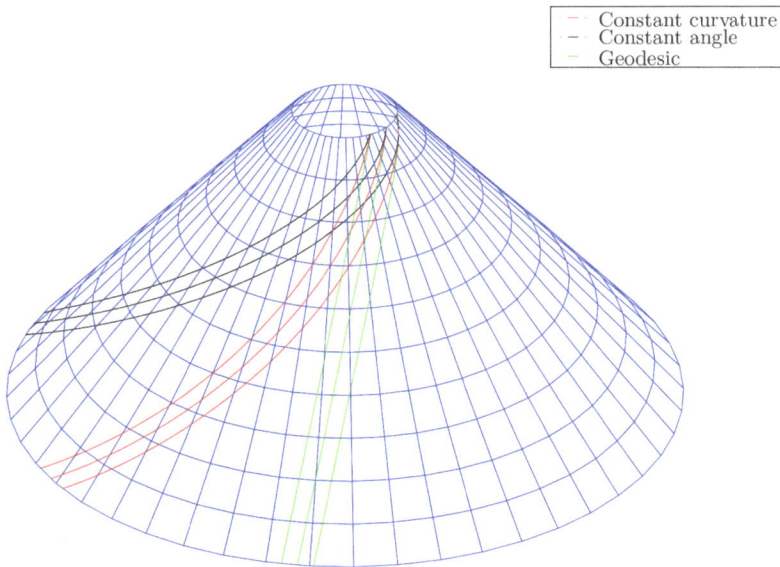

- - - Constant curvature
- -   Constant angle
- -   Geodesic

Figure 5.6 Constant curvature, geodesic, and constant angle paths on a cone with fiber orientation of the small and big radius equal to 45°: paths on the flattened cone [5.6].

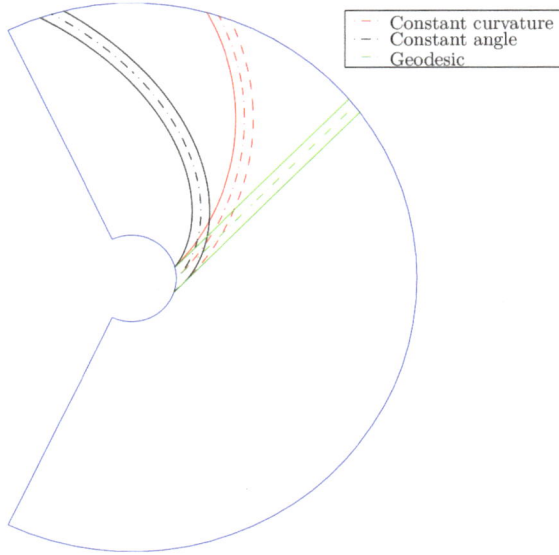

5.2.1.
Constant Angle (Rosette)

When generating fiber placement paths, there is often a desire to maintain a fiber angle within some tolerance of the rosette system. This is often one of the benefits of AFP in that the system is capable of producing more accurate fiber angles than traditional hand layup, especially on highly contoured surfaces. To achieve this outcome, a rosette, also referred to as constant angle, layup strategy is used. In general, each course generated with a rosette strategy will attempt to follow the specified ply angle relative to a defined rosette system, within some tolerance. The allowable tolerance will vary depending on many factors, such as the criticality of the structure, the ability to steer the material, and the complexity of the surface. A deviation limit of 2° is typical for simple surfaces, while a deviation of up to 5 to 10° could be used for more complex surfaces. The definition of appropriate deviation amounts will require communication between the process planner and designer. Typically, parts with severe contours or closed section parts with significant cross-sectional variation require the use of the rosette method or a combination of the rosette and parallel methods. Otherwise, fiber orientation is likely to significantly deviate from the desired value.

When generating rosette paths, it is critical to compute fiber angle deviation to assess the performance of a given course. This computation is performed with the vector defining the desired fiber angle (f_i) and the actual fiber angle (e_i). Using a dot product, the fiber deviation is expressed as follows:

$$\theta = \cos^{-1}\left(\frac{f_i \cdot e_i}{|f_i||e_i|}\right)$$

5.1

Keep in mind that, when generating these courses, it is key to keep an accurate reference for the desired fiber angle. For simple surfaces, this can be a regular axis system that applies across the entire surface. However, for more complex surfaces, it is often necessary to define multiple rosette axis systems that apply to various portions of the tool surface. Combining the deviation computation with the tolerance mentioned previously provides a range of possible solutions for a rosette path. For instance, when stepping along a surface, the next step in the path could be in any direction ±10° of the current tangent of the course. The specific solution is dependent on tangency requirements for the course centerline, steering allowances, and fiber deviation tolerance.

When using the rosette strategy on contoured surfaces, individual tows on the outer edges of the course will be either added or dropped so that the course can be steered to the appropriate orientation, while being positioned adjacent to the previous course. An input necessary in this case is the definition of the desired overlap-to-gap ratio. Typical values are full overlap, half overlap half gap, and full gap. The full overlap and gap options will allow for up to a full overlap/gap before adding/cutting a tow on the edge of a course. Similarly, the half overlap half gap option will allow up to half a tow width of either an overlap or gap. The half overlap half gap approach tends to promote the best thickness control in laminates with significant contours.

The rosette method can also be less efficient in terms of material laydown rate compared to other methods. This is because the compaction roller's velocity over the part is generally, but not always, reduced at each tow cut or add. Modern machines are capable of "on the fly" cutting without slowing down, but older machines must slow down to perform accurate cuts. The reduction in efficiency is a function of the fiber placement surface contour, which drives the need for bandwidth adjustment.

The same method to find the reference curve is presented in [5.7], but the compaction roller position is also considered. At every point of the reference curve, a tangent plane to the surface is inserted. The center of the roller is then placed on the normal to the surface calculated at each point (**Figure 5.7**). Determining the roller path location following the reference curve allows, on very complex surfaces, to avoid defects, increasing layup efficiency.

Figure 5.7 Reference curve with roller position.

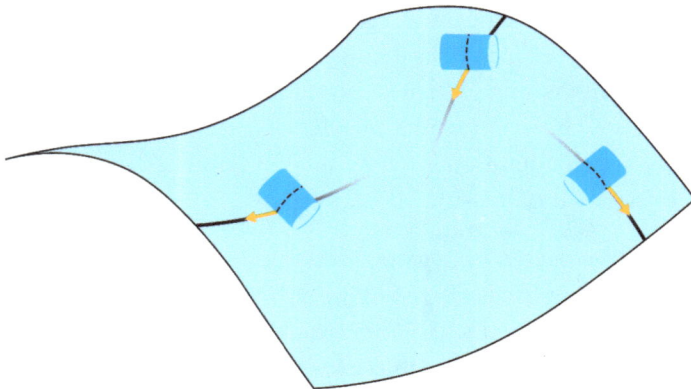

© SAE International.

5.2.2.
Geodesic (Natural) Path

These methods only focused on the fiber angle; it is possible to have a steering too severe at some point of the reference curve, making the manufacturing process difficult if not impossible. In **Figure 5.5**, the constant angle path on a conical surface is represented in black and the latter is steered on some points. This is the main reason why layup strategies need to include manufacturing configurations to generate optimal toolpaths. An effective strategy to avoid steering is to compute a geodesic guide curve, as the curvature along a geodesic path is zero. **Figure 5.5** illustrates this path on a conical shape. The geodesic path can also be known as the natural path. A geodesic is the shortest path between two points along a 3D surface in Cartesian space [5.8]. This is why on a flat panel the geodesic path is a straight line [5.9]. A geodesic path can be obtained by specifying a starting point and a direction of travel. For a general parametric surface, a geodesic path must satisfy the following system of differential equations:

$$\begin{cases} u'' + \Gamma^1_{11}u'^2 + 2\Gamma^1_{12}u'v' + \Gamma^1_{22}v'^2 = 0 \\ v'' + \Gamma^2_{11}u'^2 + 2\Gamma^2_{12}u'v' + \Gamma^2_{22}v'^2 = 0 \end{cases}$$

5.2

where Γ^i_{jk} are the Christoffel symbols. In order to solve this system of equations, four initial conditions have to be set: $u(0) = u_0$, $v(0) = v_0$, $u(1) = u_1$, and $v(1) = v_1$ for the geodesic path between two points $P_0 = S(u_0, v_0)$ and $P_1 = S(u_1, v_1)$, or $u(0) = u_0$, $v(0) = v_0$, $u'(0) = u'_0$, and $v'(0) = v'_0$ for the geodesic path starting at $P_0 = S(u_0, v_0)$ with a direction (u'_0, v'_0).

For the case of a flat surface, the system of equations in (5.2) can be simplified to obtain the parametric equation of a straight line, which is the shortest path between two points. In [5.10], a layup strategy algorithm is developed to fit a Y shape. Starting from one branch of the Y surface, and given an initial fiber angle, a geodesic path is defined. However, once at the junction of the Y, the geodesic path might change or might not be able to propagate along the surface. The different given solutions to continue the path are to go in the direction of the minimum curvature, to try and reach a geodesic path on the other branch of the Y, or to create a straight path on the other branch respecting the steering conditions for the courses.

Finally, [5.6, 5.11, 5.12] determined a reference curve for a conical shape (**Figure 5.8**). It is a good example where the parametrical approach is easier and faster than using a mesh. It is straightforward to derive the equation of a conical surface and, from that, deduce the equations for the fiber angles φ and the curvature κ of the reference curves. These two parameters are given in the following equations:

$$\sin\varphi(x) = \frac{r_0 \sin(T_0)}{r(x)} + \frac{\kappa}{\sin(\alpha)}\left(\frac{r(x)^2 - r_0^2}{2r(x)}\right) \qquad 5.3$$

$$\kappa = \left(\frac{r_1}{\bar{r}}\sin(T_1) - \frac{r_0}{\bar{r}}\sin(T_0)\right)\frac{1}{L}, \quad \left[\bar{r} = \frac{r_0 + r_1}{2}\right] \qquad 5.4$$

where

r_0, r_1, and α are the small, the big radii, and the cone angle

$r(x)$ represents the perpendicular distance from the revolution axis to a point on the shell and varies linearly for this shell configuration:
$r(x) = r_0 + x \sin\alpha$

L is the length along the surface

T_0 and T_1 are, respectively, the fiber orientation at the small and big radius of the cone

Figure 5.8 Cone geometry by Tatting et al.

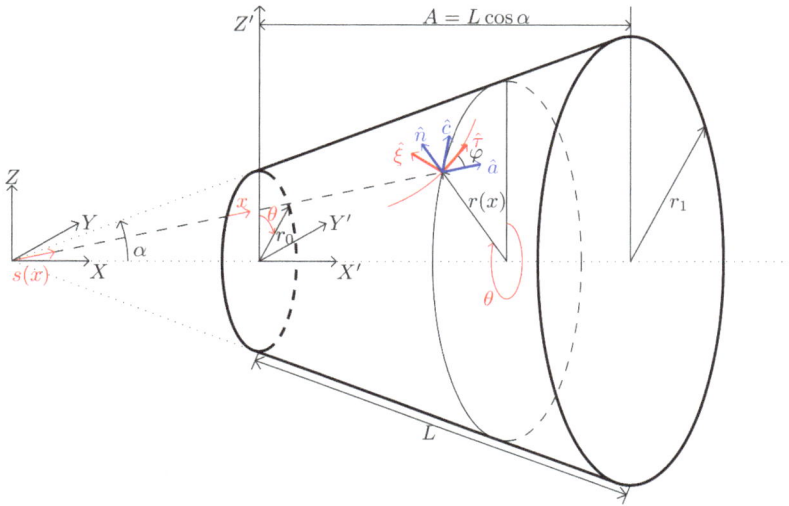

A resolution of Equations (5.2) and (5.3) keeping the fiber angle constant gives the reference curve on the conical surface with a fixed angle.

Hand Layup

Duration: 16 minutes

Description: This video, part of "Composites A-Z: 30 Days of Composites," focuses on hand layup and spray up techniques. It details hand layup as a simple, low-cost, open molding method for low-volume production, involving gel coat application, reinforcement placement, resin application, and consolidation. Spray up, also an open molding process, uses a chopper gun for faster application of resin-saturated chopped fibers, though both methods have high labor costs and moderate final part costs.

Scan the QR code below to watch the video.

5.2.3.
Variable Angle

The fiber orientation can vary along the reference curve. This is in contradiction with what is said in the section about fiber direction, as the fiber directions are not constant anymore. However, this variation of the fiber direction leads to a variable stiffness [5.13, 5.14]. The higher degree of freedom for the reference curve allows the creation of structures that account for non-unidirectional constraints. Nevertheless, the calculations and the optimizations are harder. In this section, different strategies to define the reference curve with variable angle fibers will be explained.

5.2.3.1.
Constant Curvature

For a general surface, the following system of second-order differential equations in terms of the surface parameters u and v has to be solved numerically with a prescribed geodesic curvature k_g, to obtain a constant curvature path:

$$
\begin{cases}
u'' + \Gamma_{11}^1 u'^2 + 2\Gamma_{12}^1 u'v' + \Gamma_{22}^1 v'^2 = \dfrac{k_g\left(Fu' + Gv'\right)\sqrt{Eu'^2 + 2Fu'v' + Gv'^2}}{\sqrt{EG - F^2}} \\[4mm]
v'' + \Gamma_{11}^2 u'^2 + 2\Gamma_{12}^2 u'v' + \Gamma_{22}^2 v'^2 = \dfrac{-k_g\left(Fu' + Gv'\right)\sqrt{Eu'^2 + 2Fu'v' + Gv'^2}}{\sqrt{EG - F^2}}
\end{cases}
\tag{5.5}
$$

In Equation (5.5), E, F, and G represent the first fundamental coefficients of the surface and Γ_{jk}^i represents the Christoffel symbols.

For a flat plate, a constant curvature path is an arc, with a possible parametrization:

$$
\begin{cases}
x(t) = x_0 + \dfrac{1}{k_g}\cos(t) \\[4mm]
y(t) = y_0 + \dfrac{1}{k_g}\sin(t)
\end{cases}
\tag{5.6}
$$

where

(x_0, y_0) are the coordinates of the center

$1/\kappa$ is the radius of curvature

Constant curvature paths are frequently used as trials to determine the critical radius at which wrinkling will occur for a given set of process parameters.

5.2.3.2.
Constant Curvature Path within a Flat Panel

Another polynomial description for finding reference curves using a meshed surface with a known surface equation in the x–y system (or in a polar system) is available [5.10]. Assuming that the path function is $z = f(x, y)$, one can deduce the fiber angle, θ, in each finite element center as follows:

$$z = a_1x + a_2y + a_3xy + a_4x^2 + a_5y^2 \qquad \textbf{5.7}$$

$$\theta(x,y) = \begin{cases} \tan^{-1}\left(-\dfrac{a_1 + a_3y + 2a_4x}{a_2 + a_3x + 2a_4y}\right), & a_2 + a_3x + 2a_4y \neq 0 \\[2mm] \dfrac{\pi}{2}, & a_2 + a_3x + 2a_4 \end{cases} \qquad \textbf{5.8}$$

In this method, the fiber angle is supposed to be constant in a finite element. The reference curve can then be defined as having a_1, a_2, and a_3 equal to zero, creating a constant curvature path. Detailed information is available in the referenced work.

Another strategy that implies a reference curve with a constant curvature on a flat panel is given in Gurdal and Tatting [5.9]. This strategy lies in the variation of the fiber angle between two points, each one having a different fiber angle T_0 and T_1 separated by a distance d. T_0 defines the starting point of this path. Between T_0 and T_1, a constant curvature arc with radius R^* is defined.

For a geometry such as a cone, the constant curvature can be computed by keeping κ constant in Equation (5.4). A representation of this reference curve is given in **Figure 5.5**. The case of a beam and a cylinder is studied, respectively, in Zamani, Haddadpour, and Ghazavi [5.15], and Blom, Stickler, and Gurdal [5.16] using the same method. The case of a planar surface is developed in Blom et al. [5.17], and Nik et al. [5.18] where the fiber angles follow a constant curvature path from a boundary to the center of the surface.

5.2.3.3.
Linear Variation

Another variable fiber angle layup strategy is based on the linear variation of the fiber directions along the path. This strategy lies in the linear variation of the fiber angle between two points, each one having a different fiber angle T_0 and T_1 separated by a distance d. T_0 defines the starting point of this path. The axis system of fiber orientation is defined by rotating the rosette by an angle φ. This new axis defines a new fiber orientation called r. The fiber path is then defined by $\varphi < T_0 | T_1 >$ and varies linearly along r from T_0 to T_1.

One can deduce the fiber angle, $\theta(r)$, as a function of r in the polar coordinate system:

$$\theta(r) = \begin{cases} \phi + (T_0 - T_1) \cdot \dfrac{r}{d} + T_0, & -d \leq r \leq 0 \\[2mm] \phi + (T_1 - T_0) \cdot \dfrac{r}{d} + T_0, & 0 \leq r \leq d \end{cases} \qquad \textbf{5.9}$$

The reference curve repeats indefinitely with a $2d$ period until it reaches a boundary. Further details can be found in Tatting, B. F. (2002). Design and manufacture of elastically tailored tow placed plates (NASA/CR-2002-211919). National Aeronautics and Space Administration.

5.2.3.4.
Nonlinear Variation

Nonlinear angle variations have been employed to obtain higher structural performance [5.19]. Different methods have been used to define the layup trajectories with this nonlinear variation and are explained in this section.

5.2.3.4.1.
Free Form

First, B-spline curves have been used to define a parametrical equation for reference curves. However, this method has its limits as larger amounts of control points reduce its effectiveness. This results in a low-resolution path with bad connectivity between the control points [5.20]. A Bezier curve, frequently used in computer graphics to model a smooth curve, is another way to define the reference curve with a parametrical equation of the path on the surface using a set of control points [5.21]. In [5.22], the Bezier curve is represented with a vector equation including the control points and the junction angles between each point. An example of this equation for a Bezier curve between two control points is given as follows:

$$\vec{B}(t) = (1-t)^2 \vec{P_0} + 2(1-t)t\vec{Q_1} + t^2 \vec{P_2}, \quad t \in [0,1] \qquad \text{5.10}$$

with

$$\vec{P_0} = (0,0) \qquad \text{5.11}$$

$$\vec{Q_1} = \left(\beta_1 a, \beta_1 a \tan(\alpha_0) \right) \qquad \text{5.12}$$

$$\vec{P_1} = \left(a, \beta_1 a \tan(\alpha_0) + (1 - \beta_1) a \tan(\alpha_1) \right) \qquad \text{5.13}$$

where

P_0 is the starting point

P_1 is the end point

Q_1 is the junction between the two of them

β_1 is the angle variation coefficient that defines the location of the junction point Q_1

One can then deduce the coordinates of any point on the curve as follows:

$$x = (1 - 2\beta_1)t^2 + 2\beta_1 t, \quad 0 \le x, \quad t \le 1 \qquad \text{5.14}$$

$$y = \left((1 - \beta_1) \tan(\alpha_1) - \beta_1 \tan(\alpha_0) \right) t^2 + 2\beta_1 \tan(\alpha_0) t \qquad \text{5.15}$$

It is also possible to add other segments to increase the freedom of the tow path. To do so, a new parameter needs to be introduced that makes the link between the different junction points:

$$\gamma = (i - 1 + \beta_i) / N \qquad \textbf{5.16}$$

where i and N are, respectively, the current segment number and the total number of segments.

In Parnas, Oral, and Ceyhan [5.13], the surface is considered as a bicubic Bezier surface with 16 control points as design perimeters. The control points form a mesh, and the fiber angle can be written in a cubic polynomial form:

$$s(r) = ar^3 + br^2 + cr + d \quad \text{and} \quad r = x\cos(\alpha) + y\sin(\alpha) \qquad \textbf{5.17}$$

As the fiber angle is constant in a finite element and considering the point $M(x_m, y_m)$, the center of the finite element, the fiber angle within a finite element is as follows:

$$\theta = \tan^{-1}[3a(x_m\cos(\alpha) + y_m\sin(\alpha))^2 + 2b(x_m\cos(\alpha) + y_m\sin(\alpha) + c] + \alpha \qquad \textbf{5.18}$$

In Lemaire, Zein, and Bruyneel [5.23], the reference curve trajectory is optimized following different control points. The control points are defined on the surface, and the fiber angle varies until the compliance is the lowest possible. The optimization problems are resolved thanks to finite difference sensitivities. The curvature of the surface can also be considered in the optimization to decrease the compliance.

5.2.3.4.2.
Polynomial
Multiple research studies, such as Blom, Abdalla, and Gurdal [5.19], Alhajahmad, Abdalla, and Gurdal [5.24], and Honda and Narita [5.25], use a mesh and a cubic polynomial function to determine the fiber angle along the surface:

$$f(x,y) = c_{00} + c_{10}x + c_{01}y + c_{20}x^2 + c_{11}xy + c_{02}y^2 + c_{30}x^3 + c_{21}x^2y + c_{12}xy^2 + c_{03}y^3 \qquad \textbf{5.19}$$

The different coefficients of the polynomial function can vary with the surface as they determine the surface shape. The fiber angle, θ, is also considered as constant in a finite element (but can vary from one finite element to another) so it is calculated in the center of the finite element (x_C, y_C):

$$\theta\left(x_C, y_C\right) = \tan^{-1}\left(-\frac{\partial f / \partial x}{\partial f / \partial y}\right), \quad \text{when } \partial f / \partial y = 0, \quad \theta = 90° \qquad \textbf{5.20}$$

The method used is more efficient than those using spline functions [5.20]. The fiber shape is defined using polynomial functions, while when using spline functions, simultaneous equations must be solved to achieve the same result. The path is then optimized by genetic algorithms.

In Wu et al. [5.26], Lagrangian polynomial functions are used to determine the reference curve. The following equation gives the expression of the fiber angle on the surface:

$$\theta(x, y) = \sum_{m=0}^{M-1}\sum_{n=0}^{N-1} T_{mn} \cdot \prod_{m \neq 1}\left(\frac{x - x_i}{x_m x_i}\right) \cdot \prod_{n \neq j}\left(\frac{y - y_j}{y_n - y_j}\right) \qquad \textbf{5.21}$$

where (x_i, y_j), (x_m, y_n) are the x–y coordinates of reference points. The reference curve is then obtained by resolving the equation at different reference points.

CGTech AFP Programming

Duration: 60 minutes

Description: This video, part of "Composites A-Z: 30 Days of Composites," focuses on CGTech's AFP path generation techniques. It explores NC programming for AFP, contrasting traditional methods with composites-specific approaches. Key topics include programming goals (precision, collision-free, time-efficient), considerations (feeds/speeds, surface quality, fixturing), and AFP path generation strategies (rosette, natural, limited steering, parallel), with practical steps for effective path planning and example demonstrations.

Scan the QR code below to watch the video.

5.3.
Coverage Strategies

In this section, the major coverage strategies are detailed. There are three strategies that can be used to cover the entire surface. The first one is to compute the different curves independently, and the two others are to compute all the course paths from a reference curve (**Figure 5.9**).

Figure 5.9 Projection of a major axis on the surface.

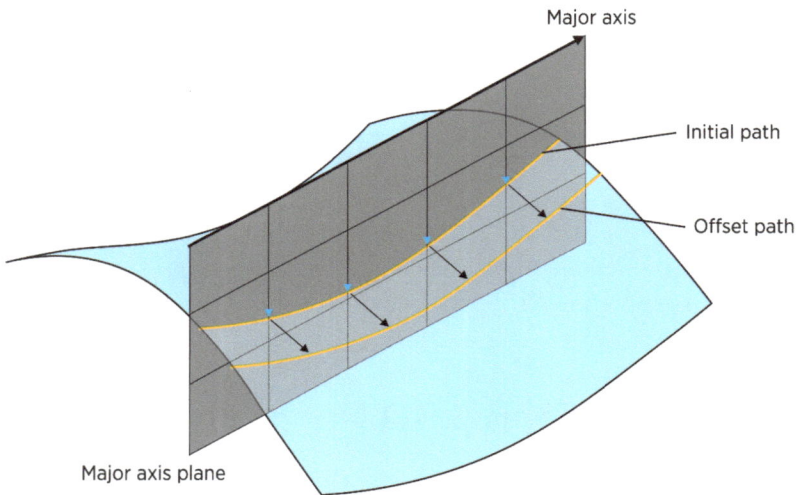

© SAE International.

5.3.1.
Parallel Curves
The parallel curve, or offset curve, strategy is the most common one for path planning on surfaces with minor curvature. Adjacent curves on the surface are computed from the reference curve to achieve total coverage. Tow paths within a course must be determined using this method due to the roller mechanism that ensures all tows within a course are parallel. However, this can lead to limitations in terms of

intra-course defect prediction for complex surfaces since parallel paths do not produce gaps or overlaps. To explain this coverage strategy, the distinction between the parametrical approach and the usage of a mesh will be investigated.

5.3.1.1.
Using a Parametrical Approach

To compute the parallel curves parametrically, a closed-form solution for continuous planar curves exists by taking equidistant points following the normal vector along the curve. This can be expressed as follows [5.4, 5.27]:

$$C(t):\begin{cases} x(t) = u_c(t) \\ y(t) = v_c(t), \\ z(t) = 0 \end{cases} \quad C_p(t):\begin{cases} x_p(t) = u_c(t) - d\dfrac{v_c'(t)}{\left(u_c'^2(t)+v_c'^2(t)\right)^{1/2}} \\ y_p(t) = v_c(t) + d\dfrac{u_c'(t)}{\left(u_c'^2(t)+v_c'^2(t)\right)^{1/2}} \\ z_p(t) = 0 \end{cases} \qquad \textbf{5.22}$$

where $C(t)$ is the reference curve and $C_p(t)$ is a parallel curve at a distance, d, from the original, and d can be either a positive number or a negative number.

For the case of a general surface, a closed-form solution for the parallel curves does not exist in most cases. Hence, several algorithms [5.7, 5.28, 5.29] have been developed to compute offset/parallel curves numerically.

For instance, a similar approach for the planar case is used in Schueler, Miller, and Hale [5.28] to find parallel curves by following the vector normal to the reference curve. This vector named O can be found by taking the cross product between the tangent vector to the curve and the normal vector to the surface.

Then, at a distance, d, along vector O, a point, P', is projected to the surface following the normal vector using a global closest technique. This process is repeated at every point step along the curve to obtain the new parallel curve. The resulting error from using this technique is reported to be [5.28]:

$$Error = d\left(1 - \frac{\psi}{\tan(\psi)}\right)$$

5.23

Therefore, the error increases by taking a further offset curve in the case of wider roller and in the case of highly curved surface.

A more accurate method is presented in Schueler, Miller, and Hale [5.28], Shirinzadeh et al. [5.29], Wang, Zhang, and Zhang [5.30], and Yan et al. [5.7] by taking the intersection between the plane perpendicular to the curve and the mold surface. To do so, a numerical approach presented in Limaiem and Trochu [5.31] is used to determine the resulting curve. Then, the offset point can be found by taking the required distance along the perpendicular arc. A last step is needed to obtain a complete offset path in the case where the reference path is shorter than the offset one that does not reach a boundary. In this case, the offset curve is completed by interpolating the last point from the calculated ones until it reaches the boundary.

Three other methods are presented in Galvez, Iglesias, and Puig-Pey [5.32] to compute parallel curves on a non-uniform rational basis spline (NURBS) surface. The first method named section curves is similar to the ones presented in Schueler, Miller, and Hale [5.28], Shirinzadeh et al. [5.29], Wang, Zhang, and Zhang [5.30], and Yan et al. [5.7]. The other two consist of generating orthogonal curves to the reference by taking either vector-field curves or geodesic curves. Once the orthogonal curve is defined in either of these methods, the offset points can be calculated at the required distance from the reference curve, and finally, the new parallel curve is obtained by interpolating these points.

The advantage of computing parallel curves is that the offset curves are equidistant so there are no gaps or overlaps between the paths or the courses during the layup process. However, considering a complex surface, the fiber directions in the offset curve can change. To optimize the fiber direction and to avoid deviation of these angles, [5.33] considers an interval of direction and tests the deviation of the offset fibers while the fiber direction in the reference path varies. The reference curve that conserves the most fiber directions in the offset path constant is then selected.

5.3.1.2.
Using a Mesh: Fast-Marching Method

This method has been introduced by Bruyneel and Zein [5.34] and Lemaire, Zein, and Bruyneel [5.23] and is based on the Eikonal equation (**Figure 5.10**). This equation is mostly used in optics to calculate the propagation of a wave with a particular speed. Hence, from a wave, one can calculate the different positions of this wave at every time once it starts propagating.

Figure 5.10 The different steps of the fast-marching method to offset a reference curve.

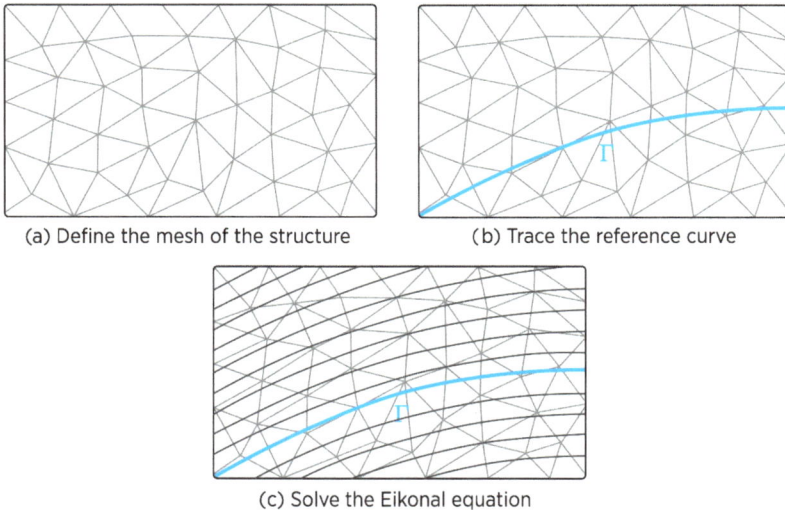

(a) Define the mesh of the structure

(b) Trace the reference curve

(c) Solve the Eikonal equation

This method starts from a random reference curve on the surface. First, the reference curve needs to be discretized. The intersection points between the mesh and the reference curve form the discretized reference path. For initialization, all these points have a time value of 0. Then, the reference curve is propagated at a defined speed so every node of the mesh will hit the propagated curve at a certain time.

Knowing the time value of two nodes of one mesh triangle, we can calculate the time value of the last node. On an accurate triangular mesh, given the time values for points A and B, T_A and T_B, one can find the time value for point C, T_C.

Using the following equations and with $1/f$, the propagation speed of the reference curve is calculated:

$$\theta = \arcsin\left(\frac{T_B - T_A}{f \cdot AB}\right)$$

5.24

$$h = BC \sin(\beta - \theta)$$

5.25

One can deduce T_C as follows:

$$T_C = h \cdot f + T_B$$

5.26

To propagate this calculation, the fast-marching method explained in Brampton, Wu, and Kim [5.35] is used. The principle of this method is to look for the neighbor nodes of the one with the lowest time value. Then, all the nodes around this one are updated. The initial node will not be considered anymore, and the update is done from the next node with the lowest value. It is important to notice that the time value of each node is updated only if the calculated value is smaller than the previous one.

Once all the node time values are known, for each time value an offset curve of the reference curve can be drawn. Knowing the speed and the time, an offset curve with the proper distance from the reference curve can be found. The offset curve joins the different iso-value points. However, to obtain a real parallel curve, the reference curve must be considered as infinite (meaning it goes through the boundaries) when propagated with the fast-marching method. If the reference path is not extended, the offset curves will not necessarily be parallel to it at every point (**Figure 5.11**).

Figure 5.11 Difference between a non-extended and an extended reference curve.

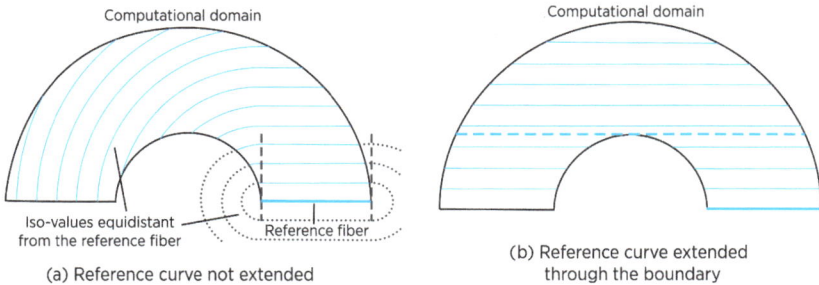

(a) Reference curve not extended

(b) Reference curve extended through the boundary

Reprinted from Computers & Structures, Michaël Bruyneel, Samih Zein, A modified Fast Marching Method for defining fiber placement trajectories over meshes, Copyright 2013, with permission from Elsevier.

Thanks to this method, equidistant curves are obtained. This method prevents gaps and overlaps, but other default behaviors linked to the steering mechanism may still occur. The direction of the fiber will be well respected if the reference curve follows the proper direction.

AFP Wind Tunnel Case Study

Duration: 31 minutes

Description: This video, part of "Composites A-Z: 30 Days of Composites," features Jacob Tury presenting a NASA Langley case study. It covers the ISAAC system, wind tunnel blade reverse engineering, AFP tool design, process planning with VCP/VCS, and manufacturing using 3D-printed prototypes, concluding with insights on complex shape fabrication.

Scan the QR code below to watch the video.

5.3.2.
Shifted Curves

In the case of shifted curves, the reference curve is simply shifted along its perpendicular direction on the surface by applying a translation. The advantage of this method is its simplicity, but the inconvenience is that, on a complex surface, the fiber directions of the offset path are not guaranteed, and the presence of gaps and overlaps is possible.

5.3.3.
Independent Curves

Another possibility to cover the entire surface is to draw the different curves independently. Regarding complex surfaces, independent curves can be a solution to limit extreme steering. To cover the surface, it is possible to draw the courses staggered one to another with a constant length and in a different direction [5.28]. If the surface is complex, the different courses are not necessarily parallel to each other if they all respect the required direction. This will induce gaps and overlaps, as shown in **Figure 5.12**.

Figure 5.12 Gaps and overlaps induced by independent direction curves.

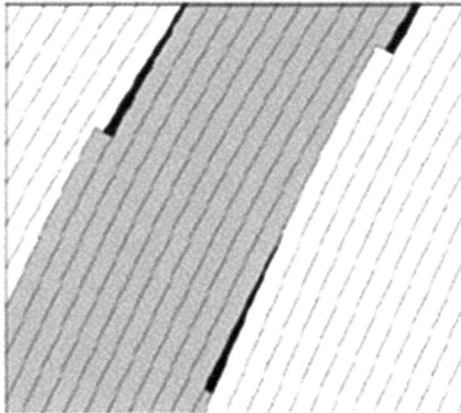

Used with permission of the American Society of
Mechanical Engineers ASME, from Approximate
Geometric Methods in Application to the
Modeling of Fiber Placed Composite Structures,
Schueler, Kurt; Miller, James; Hale, Richard, 2004

Another strategy based on the same principle involves independent fibers following the correct direction while also controlling the presence of gaps and overlaps. This approach requires many short courses. This strategy has been studied by Favaloro and Hauber [5.36] for a conical surface. The drawback of this method is that it needs a lot of short independent courses, so the structure is less resistant to constraints because of the important number of cuts.

Application

Midplane Strains and Curvatures

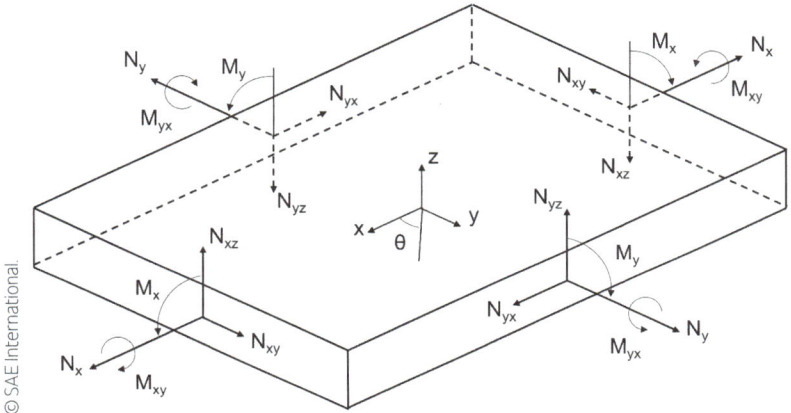

Recall that the applied forces and moments are related to the midplane strains and curvatures via the ABD matrix:

$$\begin{bmatrix} N_x \\ N_y \\ N_{xy} \\ M_x \\ M_y \\ M_{xy} \end{bmatrix} = \begin{bmatrix} A_{11} & A_{12} & A_{16} & B_{11} & B_{12} & B_{16} \\ A_{12} & A_{22} & A_{26} & B_{12} & B_{22} & B_{26} \\ A_{16} & A_{26} & A_{66} & B_{16} & B_{26} & B_{66} \\ B_{11} & B_{12} & B_{16} & D_{11} & D_{12} & D_{16} \\ B_{12} & B_{22} & B_{26} & D_{12} & D_{22} & D_{26} \\ B_{16} & B_{26} & B_{66} & D_{16} & D_{26} & D_{66} \end{bmatrix} \begin{bmatrix} \varepsilon_x^0 \\ \varepsilon_y^0 \\ \gamma_{xy}^0 \\ \kappa_x^0 \\ \kappa_y^0 \\ \kappa_{xy}^0 \end{bmatrix}$$

We will use the results of the ABD matrix that we computed previously for the [0/90/90/0] layup:

$$[ABD] = \begin{bmatrix} 1.92 * 10^9 & 5.79 * 10^7 & 0 & 0 & 0 & 0 \\ 5.79 * 10^7 & 1.92 * 10^9 & 0 & 0 & 0 & 0 \\ 0 & 0 & 1.43 * 10^8 & 0 & 0 & 0 \\ 0 & 0 & 0 & 1.07 * 10^5 & 1.93 * 10^3 & 0 \\ 0 & 0 & 0 & 1.93 * 10^3 & 2.12 * 10^4 & 0 \\ 0 & 0 & 0 & 0 & 0 & 4.78 * 10^3 \end{bmatrix}$$

We will now use forces and moments to compute strains and curvatures. To accomplish this, we must solve for the midplane strains and curvatures. This gives us the equation:

$$
\begin{bmatrix} \varepsilon_x^0 \\ \varepsilon_y^0 \\ \gamma_{xy}^0 \\ \kappa_x^0 \\ \kappa_y^0 \\ \kappa_{xy}^0 \end{bmatrix}
=
\begin{bmatrix}
a_{11} & a_{12} & a_{16} & b_{11} & b_{12} & b_{16} \\
a_{12} & a_{22} & a_{26} & b_{12} & b_{22} & b_{26} \\
a_{16} & a_{26} & a_{66} & b_{16} & b_{26} & b_{66} \\
b_{11} & b_{12} & b_{16} & d_{11} & d_{12} & d_{16} \\
b_{12} & b_{22} & b_{26} & d_{12} & d_{22} & d_{26} \\
b_{16} & b_{26} & b_{66} & d_{16} & d_{26} & d_{66}
\end{bmatrix}
\begin{bmatrix} N_x \\ N_y \\ N_{xy} \\ M_x \\ M_y \\ M_{xy} \end{bmatrix}
$$

Here, the lowercase a, b, and d represent the inverse of the ABD matrix. We will apply an axial force of $N_x = 1000$ N/m, resulting in strains and curvatures of

$$
\begin{bmatrix} \varepsilon_x^0 \\ \varepsilon_y^0 \\ \gamma_{xy}^0 \\ \kappa_x^0 \\ \kappa_y^0 \\ \kappa_{xy}^0 \end{bmatrix}
=
\begin{bmatrix}
5.21 * 10^{-7} \\
-1.57 * 10^{-8} \\
0 \\
0 \\
0 \\
0
\end{bmatrix}
$$

If we do the same exercise for a non-symmetric laminate such as a cross-ply laminate with the stacking sequence [0/90/0/90], we get the strains and curvatures of

$$
\begin{bmatrix} \varepsilon_x^0 \\ \varepsilon_y^0 \\ \gamma_{xy}^0 \\ \kappa_x^0 \\ \kappa_y^0 \\ \kappa_{xy}^0 \end{bmatrix}
=
\begin{bmatrix}
6.12 * 10^{-7} \\
-1.85 * 10^{-8} \\
0 \\
4.10 * 10^{-5} \\
0 \\
0
\end{bmatrix}
$$

Now, we got an interesting result. Our axial force has now coupled with bending and caused a slight curvature. This is due to the asymmetric laminate and a non-zero B matrix. This is a critical parameter to look out for when designing composite laminates. In industry, the standard often used is to have a balanced and symmetric laminate.

5.3.4.
Hybrid Methods

Any offset to a reference curve is likely to force a deviation from desired attributes such as fiber angle or steering radii. Therefore, for optimal material placement, the best choice is often to combine reference curve generation and coverage strategies into a hybrid method. Such a method allows for an initial reference curve to be generated and offset across the surface. The offset curve can then be assessed to determine whether it violates given thresholds, such as fiber angle deviation. In the case that the offset curve is in violation, a new reference curve is computed, and the offset strategy begins again from this new reference curve. In general, a process planner can use any combination of reference curves and offset strategies to create a hybrid method. However, in practice, there are generally three commonly used hybrid methods. These are (1) a combination of rosette reference curves and parallel offsets with a threshold criterion associated with the fiber angle, (2) a combination of natural reference curves and parallel offsets with a threshold criterion associated with the steering angle, and (3) a combination of natural reference curves and parallel offsets with a threshold criterion associated with the fiber angle.

The rosette-parallel method begins with a rosette reference curve computed based on the starting point location. The reference curve is then offset using the parallel strategy, but after each parallel course is generated, all its tows will be checked for any violation of the tow direction tolerance. If there is a violation of the allowed angle deviation, the path will be replaced by a new reference curve that follows the rosette method. Subsequent courses will then be created parallel to this new reference curve instead of the original one until a tow's direction departs too far from the ply angle again. Typically, this hybrid method strikes the best balance between fiber angle deviation and the occurrence of gaps and overlaps. For highly contoured surfaces, this hybrid strategy can result in a complete rosette strategy because the parallel paths will almost always violate the fiber angle deviation limit.

Both the natural-parallel methods behave similarly, with the main difference being the attribute that is considered to determine whether an offset curve is adequate. The initial reference curve will be generated using a natural strategy using the defined starting point. Subsequent courses will be parallel to the prior one, but after each is generated all of its tows will be checked for any violation of either the fiber angle or the steering radius. If a violation occurs, a new reference curve will be generated using the natural method, and the parallel offset curves will then be generated from the new curve. Note that it is often worth averaging the attribute being assessed over a given length of course or tow. This avoids replacing parallel courses in which a very short segment of a tow or tows violates the given threshold.

5.3.4.1.
Dynamic Layup Strategies

One could also create a highly customized hybrid strategy in which the reference curve and coverage strategies change on a point-by-point basis across the entire surface. Such a strategy is deemed the name "dynamic layup strategies" [5.37]. The dynamic layup strategy intends to abstract away the decision-making present with layup strategy selection. As previously discussed, fiber coverage methods and fiber defects are closely coupled and present a series of tradeoffs. Furthermore, the selection of a given layup strategy will provide the specified benefits when some aspects of a different layup strategy may also be desired. Thus, the current paradigm with layup strategy selection is a very discrete one. The dynamic layup strategy creates a continuous integration of multiple layup strategies, where each may be partially implemented to gain a tailored blend of performance. Additionally, performance can vary over the tool surface, such that specified layup properties can be targeted for different regions of the structure.

The dynamic layup strategy utilizes the fundamental logic from each of the previously detailed layup and coverage algorithms, enabling the ability to continuously target the minimization of various fiber defect types. Just as with the other methods, the definition of the initial curve begins with the starting point and fiber angle. This, combined with the surface normal, creates a reference frame that can be "stepped" across the surface to generate the path. The stepping process for the initial curve is referred to as a point stepper. The point stepper begins at the starting point and provides the logic that is utilized to discretely update the fiber path, which is currently under construction. From that seed point, the point stepper logic is extended outward until the ply boundaries are reached. Fundamentally, the point stepper provides the local tangent of the path, where the tangent is utilized to generate the subsequent point of the path in addition to re-evaluating the local path's tangent. To build a final course centerline, the individual iterations of the point stepper generate discrete pairs of points and path tangents. A spline curve fitting is performed with these data to generate a final continuous form of the course centerline for further processing.

The coverage strategy for this dynamic method utilizes the fundamental properties of the parallel propagation to minimize the presence of course-to-course gaps and overlaps. A new path should be parallel to the previous course to minimize gaps and overlaps. Therefore, any deviation from that parallel path can be understood as the generation of gaps and overlaps. Thus, the coverage-based point stepper is primarily concerned with minimizing deviation of the constructed path from the idealized parallel offset from the previous course. When initializing the coverage point stepper for the generation of a new course centerline, it begins when generating the idealized parallel offset from the previous course.

During each update with the coverage point stepper, the current leading point of the path is projected to the parallel reference curve. To minimize gap and overlap presence, the step direction is set toward the parallel reference curve. The total change in direction is limited, however, to avoid the creation of excess angle deviation or steering radii. Therefore, the step direction is rotated to point toward the reference curve, and the maximum amount of rotation is the allowable maximum angle deviation. The use of the maximum angle deviation for redirecting the step direction acts to minimize the amount of induced angle deviation encountered when attempting to maintain the trajectory along the parallel reference curve. As the stepping progresses, the user can essentially combine various layup strategies to create an optimal path. For example, **Figure 5.13** shows the steps associated with reducing individual defects. The final computation of the next point is then the summation of these vectors incorporating coefficients for the importance of each defect. Defect importance can even change from point to point depending on local geometry or critical structural locations. This process provides the optimal paths for every point on a given surface.

Figure 5.13 Computation of various layup strategy propagation directions.

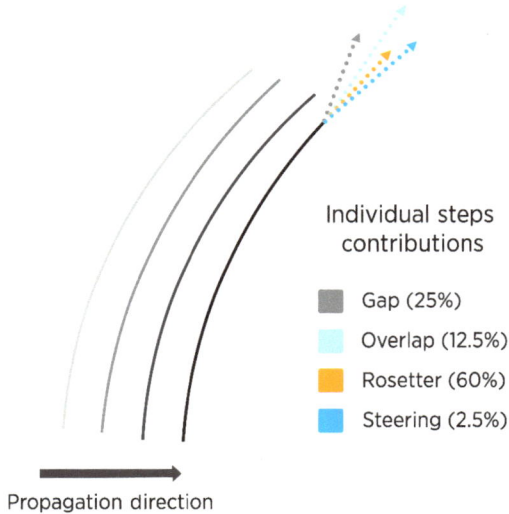

Individual steps contributions

- Gap (25%)
- Overlap (12.5%)
- Rosetter (60%)
- Steering (2.5%)

Propagation direction

Joshua Allen Halbritter, 2023.

5.4.
Through-Thickness Staggering

At this stage, we have covered the different methods we can optimize and select the most suitable layup strategies for a specific ply in the laminate. Industrial practices typically stagger coverage of plies with identical orientations to avoid the unintentional stacking of defects and the potential implications of such. To better illustrate, let us consider the example of a coverage strategy that would result in a gap at a specific location in the laminate. If the subsequent ply has a similar fiber orientation, it will result in an additional gap on top of the first one.

This would double the gap volume, creating a bigger resin-rich region. Implications can be detrimental to the structural integrity of the intended part. Typical industrial practices stagger plies based on the width of the tow. For a set fiber orientation, half the width of a tow is used as a dimension to move the starting point. This section should be expanded beyond simply addressing through-thickness staggering. Recent work by Nishan Patel highlights the significant potential of comprehensive through-thickness optimization for various types of defects through laminate-level design.

5.5.
Discussion and Summary

Determining the optimal layup toolpath for composite manufacturing with AFP is a complex yet crucial task, as it directly influences the final product's quality, performance, and overall production efficiency. A well-optimized toolpath can yield several significant benefits:

1. **Minimized Defects:** By carefully planning the fiber placement sequence and direction, manufacturers can reduce the potential for defects like wrinkles, gaps, or overlaps, leading to a decrease in costly inspection and repair times.

2. **Reduced Production Time:** Optimizing the AFP head's movements, particularly the non-productive "off-part" motion, can significantly increase the actual layup time, thereby improving overall production throughput.

3. **Enhanced Quality Control:** The chosen toolpath strategy can be rigorously validated through simulations and comparisons with safety margins to ensure accuracy and conformity to design specifications, minimizing the risk of costly errors during manufacturing.

4. **Improved Structural Integrity:** A well-optimized toolpath can lead to a more uniform distribution of fibers and resin, enhancing the composite's structural integrity and mechanical properties. Making the "as-manufactured" part close to the "as-designed" one.

5. Cost Reduction: By minimizing defects, reducing production time, and optimizing material usage, an optimized toolpath can contribute to significant cost savings throughout the manufacturing process.

In summary, while finding the optimal layup toolpath for AFP composite manufacturing is undeniably challenging, the potential rewards in terms of product quality, production efficiency, and cost-effectiveness make it an essential step in the pursuit of high-performance composite structures.

References

5.1. Rousseau, G., Wehbe, R., Halbritter, J., and Harik, R., "Automated Fiber Placement Path Planning: A State-of-the-Art Review," *Computer-Aided Design* 16, no. 2 (2019): 172-203.

5.2. Shirinzadeh, B., Alici, G., Foong, C.W., and Cassidy, G., "Fabrication Process of Open Surfaces by Robotic Fibre Placement," *Robotics and Computer-Integrated Manufacturing* 20, no. 1 (2004): 17-28.

5.3. Chen, H., Fuhlbrigge, T., and Li, X., "A Review of CAD-Based Robot Path Planning for Spray Painting," *Industrial Robot: An International Journal* 36, no. 1 (2009): 45-50.

5.4. Wehbe, R., "Modeling of Tow Wrinkling in Automated Fiber Placement Based on Geometrical Considerations," University of South Carolina, 2017, Retrieved from https://scholarcommons.sc.edu/etd/4449?utm_source=scholarcommons. sc.edu%2Fetd%2F4449&utm_medium=PDF&utm_campaign=PDFCoverPages.

5.5. Lukaszewicz, D., Ward, C., and Potter, K., "The Engineering Aspects of Automated Prepreg Layup: History, Present and Future," *Composites Part B: Engineering* 43, no. 3 (2012): 997-1009.

5.6. Blom, A., Tatting, B., Hol, J., and Gurdal, Z., "Fiber Path Definitions for Elastically Tailored Conical Shells," *Composites Part B: Engineering* 40, no. 1 (2009): 77-84.

5.7. Yan, L., Chen, Z.C., Shi, Y., and Mo, R., "An Accurate Approach to Roller Path Generation for Robotic Fibre Placement of Free-Form Surface Composites," *Robotics and Computer-Integrated Manufacturing* 30, no. 3 (2014): 277-286.

5.8. Patrikalakis, N. and Maekawa, T., *Shape Interrogation for Computer Aided Design and Manufacturing* (Berlin: Springer & Business Media, 2001).

5.9. Gurdal, Z. and Tatting, B., "Tow-Placement Technology and Fabrication Issues for Laminated Composite Structures," in *46th AIAA/ASME/AHS/ASC Structures, Structural Dynamics and Materials Conference*, Austin, 2005.

5.10. Hely, C., Birglen, L., and Xie, W.-F., "Feasibility Study of Robotic Fibre Placement on Intersecting Multi-Axial Revolution Surfaces," *Robotic and Computer-Integrated Manufacturing* 48 (2017): 73-79.

5.11. Blom, A., "Structural Performance of Fiber-Placed, Variable Stiffness Composite Conical and Cylindrical Shells," Doctoral thesis, Delft University of Technology, 2010.

5.12. Blom, A., Setoodeh, S., Hol, J., and Gurdal, Z., "Design of Variable-Stiffness Conical Shells for Maximum Fundamental Eigenfrequency," *Computers & Structures* 86, no. 9 (2008): 870-878.

5.13. Parnas, L., Oral, S., and Ceyhan, U., "Optimum Design of Composite Structures with Curved Fiber Courses," *Composites Science and Technology* 63, no. 7 (2003): 1071-1082.

5.14. Sabido, A., Bahamonde, L., Harik, R., and van Tooren, M., "Maturity Assessment of the Laminate Variable Stiffness Design Process," *Composite Structures* 160 (2017): 804-812.

5.15. Zamani, Z., Haddadpour, H., and Ghazavi, M., "Curvilinear Fiber Optimization Tools for Design Thin Walled Beams," *Thin-Walled Structures* 49, no. 3 (2011): 448-454.

5.16. Blom, A., Stickler, P., and Gurdal, Z., "Optimization of a Composite Cylinder under Bending by Tailoring Stiffness Properties in Circumferential Direction," *Composites Part B: Engineering* 41, no. 2 (2010): 157-165.

5.17. Blom, A., Lopes, C., Kromwijk, P., Gurdal, Z. et al., "A Theoretical Model to Study the Influence of Tow-Drop Areas on the Stiffness and Strength of Variable-Stiffness Laminates," *Journal of Composite Materials* 43, no. 5 (2009): 403-425.

5.18. Nik, M.A., Fayazbakhsh, K., Pasini, D., and Lessard, L., "Optimization of Variable Stiffness Composites with Embedded Defects Induced be Automated Fiber Placement," *Composite Structures* 107 (2014): 160-166.

5.19. Blom, A., Abdalla, M., and Gurdal, Z., "Optimization of Course Locations in Fiber-Placed Panels for General Fiber Angle Distributions," *Composites Science and Technology* 70, no. 4 (2010): 564-570.

5.20. Honda, S., Narita, Y., and Sasaki, K., "Maximizing the Fundamental Frequency of Laminated Composite Plates with Optimally Shaped Curvilinear Fibers," *Journal of System Design and Dynamics* 3, no. 6 (2009): 867-876.

5.21. Farin, G., *Curves and Surfaces for Computer-Aided Geometric Design* (Cambridge, MA: Academic Press, 1993).

5.22. Kim, B.C., Potter, K., and Weaver, P., "Continuous Tow Shearing for Manufacturing Variable Angle Tow Composites," *Composites Part A: Applied Science and Manufacturing* 43, no. 8 (2012): 1347-1356.

5.23. Lemaire, E., Zein, S., and Bruyneel, M., "Optimization of Composite Structures with Curved Fiber Trajectories," *Composite Structures* 131 (2015): 895-904.

5.24. Alhajahmad, A., Abdalla, M., and Gurdal, Z., "Design Tailoring for Pressure Pillowing Using Tow-Placed Steered Fibers," *Journal of Aircraft* 45, no. 2 (2012): 630.

5.25. Honda, S. and Narita, Y., "Vibration Design of Laminated Fibrous Composite Plates with Local Anisotropy Induced by Short Fibers and Curvilinear Fibers," *Composite Structures* 93, no. 2 (2011): 902-910.

5.26. Wu, Z., Weaver, P., Raju, G., and Kim, B., "Buckling Analysis and Optimisation of Variable Angle Tow Composite Plates," *Thin-Walled Structures* 60 (2012): 163-172.

5.27. Wehbe, R., Tatting, B., Harik, R., Gurdal, Z. et al., "Tow-Path Based Modeling of Wrinkling during the Automated Fiber Placement Process," in *CAMX 2017*, Orlando, 2017.

5.28. Schueler, K., Miller, J., and Hale, R., "Approximate Geometric Methods in Application to the Modeling of Fiber Placement Composite Structures," *Journal of Computing and Information Science in Engineering* 4, no. 3 (2004): 251-256.

5.29. Shirinzadeh, B., Cassidy, G., Oetomo, D., Alici, G. et al., "Trajectory Generation for Open-Contoured Structures in Robot Fibre Placement," *Robotics and Computer-Integrated Manufacturing* 23, no. 4 (2007): 380-394.

5.30. Wang, X., Zhang, W., and Zhang, L., "Intersection of a Ruled Surface with a Free-Form Surface," *Numerical Algorithms* 46 (2007): 85-100.

5.31. Limaiem, A. and Trochu, F., "Geometric Algorithms for the Intersection of Curves and Surfaces," *Computers & Graphics* 19, no. 3 (1995): 193-403.

5.32. Galvez, A., Iglesias, A., and Puig-Pey, J., "Computing Parallel Curves on Parametric Surfaces," *Applied Mathematical Modelling* 38, no. 9-10 (2014): 2398-2413.

5.33. Debout, P., "Calcul de trejets de depose dans le cadre de la fafrication de pieces aeronautiques," Doctoral dissertation, Universite Blaise Pascal-Clermont-Ferrand II, 2010.

5.34. Bruyneel, M. and Zein, S., "A Modified Fast Marching Method for Defining Fiber Placement Trajectories over Meshes," *Computers and Structures* 125 (2013): 45-52.

5.35. Brampton, C., Wu, K., and Kim, H., "New Optimization Method for Steered Fiber Composites Using the Level Set Method," *Structural and Multidisciplinary Optimization* 52, no. 3 (2015): 493-505.

5.36. Favaloro, M. and Hauber, D., "Process and Design Considerations for the Automated Fiber Placement Process," SAMPE, Cincinnati, 2007.

5.37. Halbritter, J., "Leveraging Automated Fiber Placement Computer Aided Process Planning Framework for Defect Validation and Dynamic Layup Strategies," Doctoral dissertation, University of South Carolina, 2023.

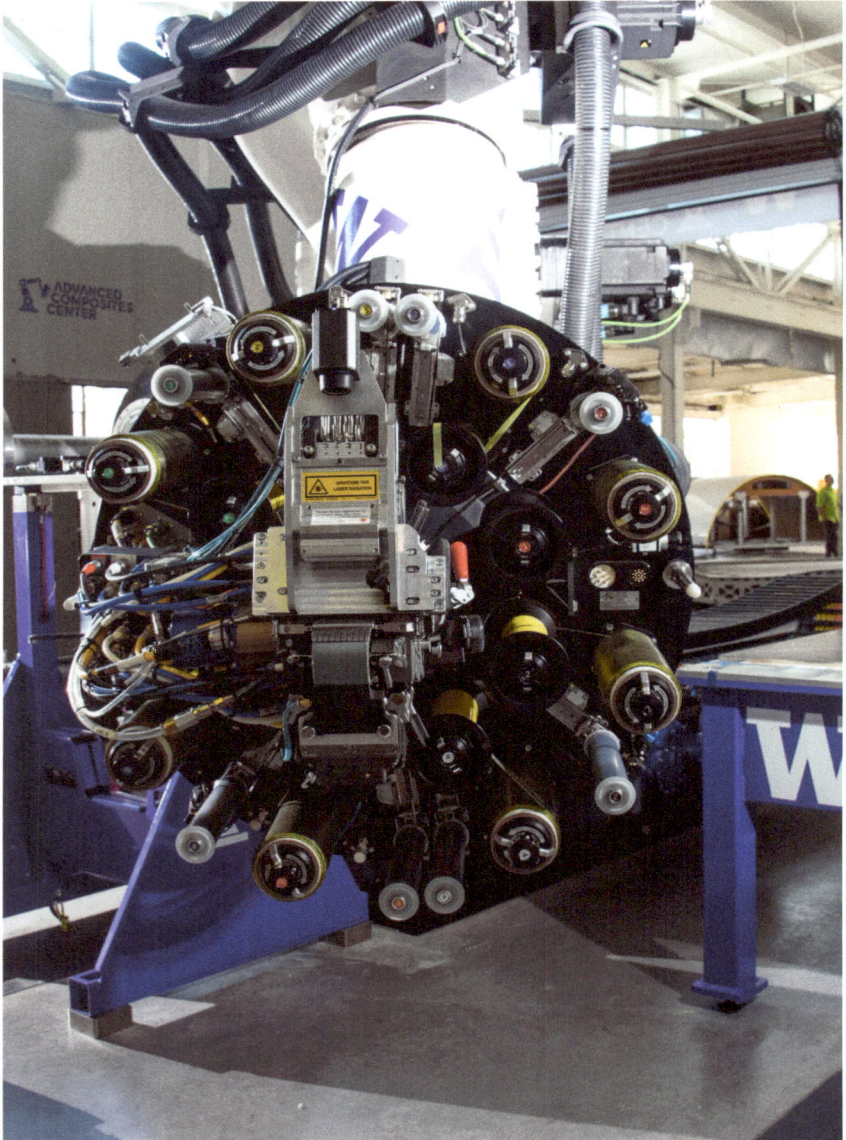

An Automated Fiber Placement machine at the University of Washington's Advanced Composites Center in Seattle.

Inspection

The quality control during AFP processes is either manually done by visual inspection of the operator [6.1] or by using various types of automated inspection methods. Due to the low contrast between the substrate and incoming tows, visual identification of defects has proven to be difficult. However, thermal imaging, laser profiling, eddy current inspection [6.2], and other non-destructive testing (NDT) techniques have been employed to ease the difficulty of inspection. The development of abilities to rapidly monitor and inspect AFP-manufactured plies is a top priority in improving manufacturing efficiency [6.3]. The current industry standard for inspection is primarily visual/manual. While frequently accurate, manual inspection is typically very time-intensive, requires expert knowledge, and reduces traceability in determining the quality of layup. The time cost of manual inspection is significant [6.4-6.6], with estimates placing the percentage of machine time dedicated to laydown being as low as 24%. This was almost entirely due to manual inspection and rework in early AFP days, with inspection time growing with the size of each part. This makes producing large-scale composites increasingly time and cost prohibitive.

6.1.
Inspection Techniques

6.1.1.
Visual

Visual inspection continues to be a widely used practice for inspecting AFP-manufactured parts. An operator will go into the cell after a ply is placed and visually look for any defects that may have been inadvertently placed during the layup. However, it can be extremely difficult to see defects due to the dark complexion of AFP materials. Therefore, visual inspection is often aided by some form of automated inspection as detailed in the following sections. The use of additional sensors outside of the human eye allows for much finer inspection resolution and increases the reliability of detecting and reworking any defects within a structure.

6.1.2.
Profilometry

Profilometry utilizes laser projections onto a surface to infer surface features from pattern deviations [6.7]. This method enables rapid profiling of a surface without considering the surface contrast (**Figure 6.1**) [6.8]. However, depending on the specific profilometer used, material type can have a direct effect on the quality of data gathered and, therefore, the accuracy of defect identification and classification [6.9]. For example, very thin materials may not have significant height changes caused by defects, therefore making them difficult to see in the profilometry data. Materials that would interfere with the laser projections, such as highly reflective materials, would also result in poor data quality. Cemenska et al. [6.4] showed that profilometers can detect gaps, overlaps, FOD, bridging, puckering, delamination, and tow twist given sufficient size. The feature recognition necessary for detecting these defects requires processing the raw data with custom algorithms that can be highly complex requiring high-performance computing for higher feed rates making it difficult for real-time feedback. It should be noted that when properly optimized, profilometry offers unrivaled surface detail, allowing for more accurate analysis and the identification of a broader range of defects. However, variation in the optical properties (e.g., refractive index, dispersion, and absorption) across materials makes this difficult. Two recent industrial in situ inspection systems offered from Coriolis and EI are both profilometry based [6.10, 6.11].

Figure 6.1 Visualization of profilometry data [6.8].

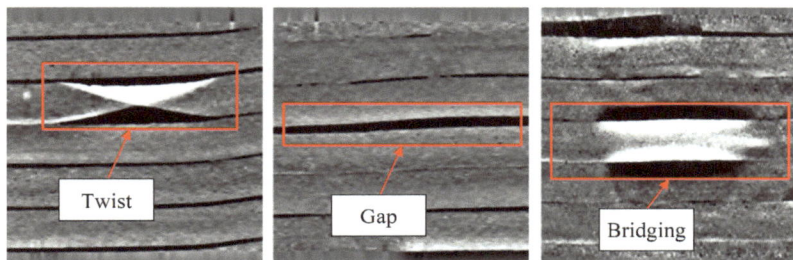

C. Sacco, A. Baz Radwan, A. Anderson and R. Harik, "NDE Inspection of AFP Manufactured Cylinders Using an Intelligent Segmentation Algorithm," in SAMPE Conference & Exhibition, Orlando, Florida, 21-24 September 2020. Reprinted by permission from the Society for the Advancement of Material and Process Engineering (SAMPE).

6.1.3.
Thermal

Thermographic monitoring is based on a thermal camera combined with (process-depending) image processing that can analyze the visible temperature difference between the laid tow and the substrate [6.12]. The temperature difference in the tow is triggered by the propagation of heat through the laminate with the heat being provided by the heat source on the AFP head. Accurately measuring the temperature profiles requires the thermal camera to be mounted onto the AFP head to measure the temperatures directly after the compaction roller. This information can be used to derive and store both tow positions and process-relevant defects. Gregory and Juarez [6.13] demonstrated that the changes in temperature profiles seen in in situ thermographic monitoring are sufficient for identifying all types of defects. Schmidt et al. [6.14] also used this technique combined with an edge detection algorithm to localize tows and detect temperature discrepancies. Juarez and Gregory [6.15] provide a depiction (**Figure 6.2**) of how defects contribute to the temperature profile. Note that the heat flow shown in this figure is from convective heat transfer of the heat energy applied to the material by the AFP head and released into the air. The temperature measurement from a gap is higher due to a smaller substrate thickness to absorb the heat and vice versa for an overlap. Various other defects such as twists, splices, and folds can be deduced from the temperature profiles.

Figure 6.2 A depiction of the effect of defects on the temperature profile [6.15].

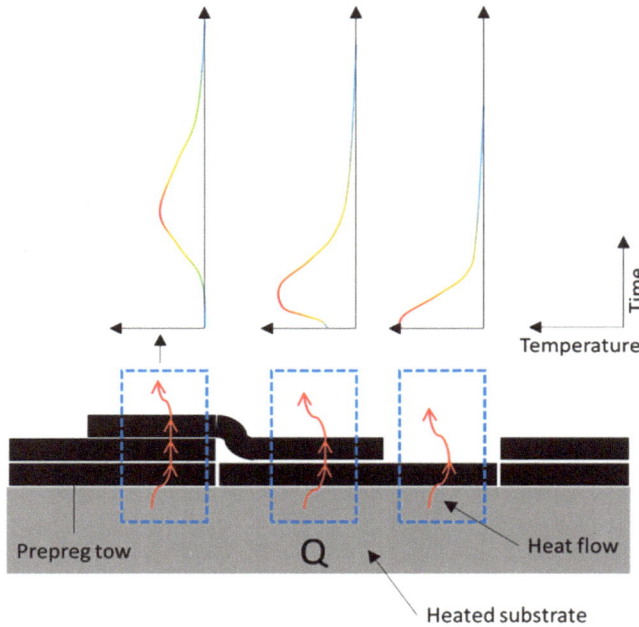

Reprinted from In situ thermal inspection of automated fiberplacement manufacturing, Juarez, Peter D.; Gregory, Elizabeth D., 2019, with the permission of AIP Publishing

6.2.
In-Process Inspection

A major distinction that must be made in the inspection and quality monitoring of AFP parts is between inspection systems that are ply-by-ply and inspection systems that are in situ. Ply-by-ply or static systems wait until the layup process or a part of it is complete and then inspect while the AFP machine is inactive. An example of such a system by Ingersoll Machine Tools is shown in **Figure 6.3**.

Figure 6.3 Ingersoll Machine Tools Automated Composite Structures Inspection System (ACSIS).

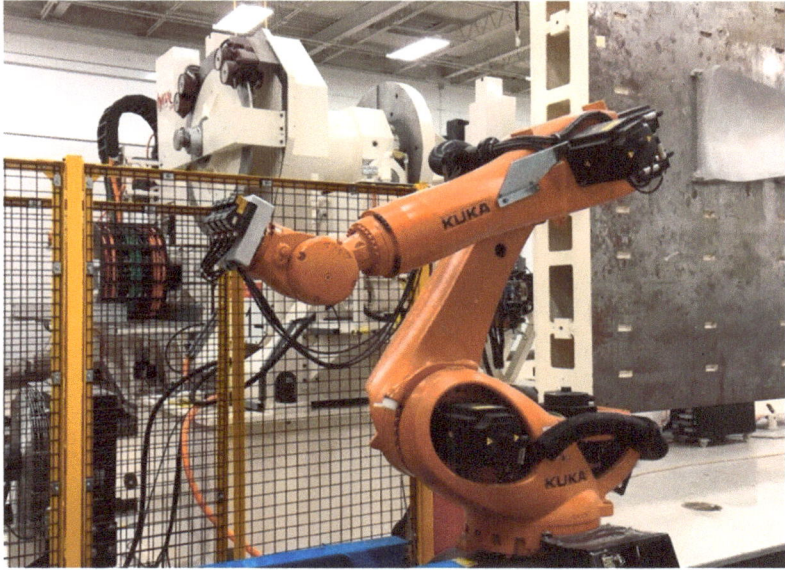

In situ methods are capable of inspecting while an AFP machine is performing layup using an attached sensor, as shown in **Figure 6.4**. This maximizes machine usage and reduces the amount of necessary machine downtime, therefore contributing to higher process throughput. The advantages of in situ systems are driving the majority of the development in AFP inspection toward this area.

Figure 6.4 Schematic of inspection system on an AFP head.

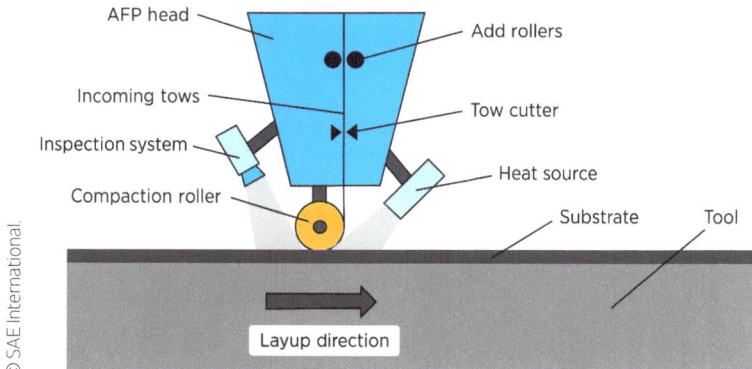

Did You Know?

Sauce Reques/Shutterstock.com

One of the cool applications of composite parts that are mostly manually made and not using AFP is musical instruments. Pictured here is a Mezzo-Forte Carbon Fiber premium line violin that is sold for $5000! The usage of composites is often attributed to comparable sounds to wooden instruments while having far greater presence, projection, and ease of playing.

The thermographic approaches to inspection mentioned previously in this chapter are all implemented as in situ systems, using the heat emanating from the layup process as a way to produce images. In addition, companies such as FlightAware [6.16] have developed in situ AFP inspection systems based on profilometry and laser line scanning. EI's RIPITx system (Real Time in Process AFP Inspection) has recently become an industrially available inspection option for profilometry-based in situ inspection [6.11].

6.2.1.
Automated Defect Detection

Once the data acquisition component of an AFP system is created, it is also often advantageous to engage it to a data analysis system to automatically identify defects through the data. Approaches to data analysis vary across applications, but a consistent class of algorithms has coalesced under the branch of computer vision and deep learning. Nearly all of the data acquisition systems that have been developed present their data in a visual manner. This design choice allows engineers to leverage advances in computer vision to perform automated defect identification.

In particular, the advent of deep learning image processing algorithms spurred on by the success of convolutional neural networks [6.17] has allowed for significant improvements in the performance of industrial inspection systems [6.18, 6.19]. Iteration on neural network architectures for computer vision has produced a number of models that achieve human or near-human levels of performance [6.20]. Recent advancements in ML and AI have resulted in computer vision systems that are significantly better than human levels. This has yet to be fully deployed in the composite manufacturing sector for production inspection systems, but the technology is expected to make an appearance in the future.

A considerable drawback to these approaches has been the large amounts of data required to properly train one of these deep object detection algorithms. New areas of emphasis are on reducing this obstacle by utilizing principles such as transfer learning [6.21, 6.22], whereby lower-level features learned on one dataset are preserved and reused on a similar dataset. In the area of AFP inspection, this may present a potential path forward for training a model to detect across multiple material types or with multiple sensors as input.

To perform a prediction or training cycle with these models, it is important to leverage many of the modern advances in computer hardware that allow for massive parallelization of neural networks. Computer hardware such as graphical processing units (GPUs) [6.17, 6.23] and field programmable gate arrays (FPGAs) [6.24-6.28] utilize parallel processing to massively increase the speed at which many neural networks run. In the case of FPGAs, this is the result of writing dedicated hardware in silicon to perform operations rather than relying on standard general-purpose processors.

Some work has begun to combine these principles to create automated detection algorithms for AFP defects. Several studies by the authors have leveraged ML, integrated systems, and advanced computer hardware to create a computer vision system for learning and identifying AFP defects [6.9, 6.29-6.31]. **Figure 6.5** shows the results of an intelligent segmentation algorithm that is capable of identifying narrow gaps and overlaps in profilometry scans of a layup.

Figure 6.5 ML segmentation of an AFP part scan for defect identification [6.30].

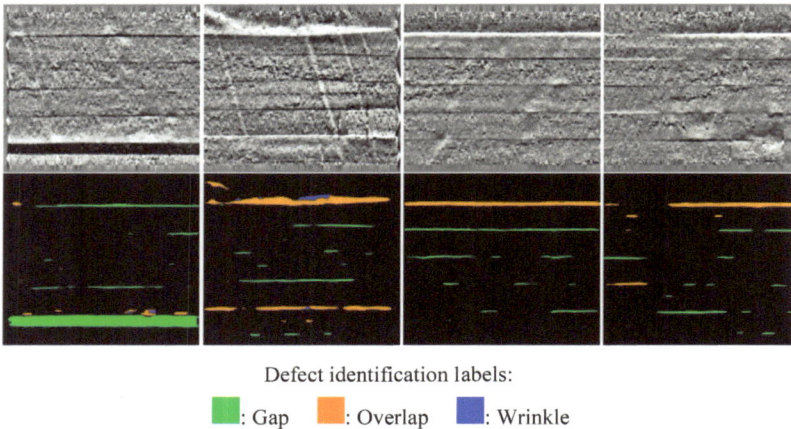

Defect identification labels:

■: Gap   ■: Overlap   ■: Wrinkle

C. Sacco, A. Baz Radwan, A. Anderson and R. Harik, "NDE Inspection of AFP Manufactured Cylinders Using an Intelligent Segmentation Algorithm," in SAMPE Conference & Exhibition, Orlando, Florida, 21-24 September 2020. Reprinted by permission from the Society for the Advancement of Material and Process Engineering (SAMPE).

6.2.2.
Multi-Robot Inspection System

Advancements in robotics have also brought about the possibility for multi-robot systems. Such a system could be a groundbreaking approach to composite manufacturing. For instance, a dual robot system could be specifically engineered for the AFP process. This system would harness the power of two specialized robotic arms, each equipped with state-of-the-art sensors and precise control mechanisms, to revolutionize the way composite structures are fabricated.

The primary robot would be the AFP machine dedicated to the meticulous task of laying down material onto the tool surface or substrate. Simultaneously, the secondary robot operates in tandem with the AFP machine, performing real-time inspections throughout the manufacturing process. Equipped with an array of advanced sensors that could include IR cameras for thermography, ultrasonic transducers for subsurface scanning, high-resolution optical systems, and acoustic emission (AE) sensors, the inspection robot meticulously monitors the integrity of the composite structure as it takes shape. This comprehensive suite of sensors enables the robot to detect and identify defects such as voids, delamination, fiber misalignment, or foreign inclusions with unparalleled accuracy throughout the entirety of the manufacturing process.

The integration of the dual robot system allows for immediate feedback loops, enabling adaptive adjustments to optimize quality and minimize defects during production. Theoretically, the data streams of the robots could connect to perform the suggested real time feedback. For instance, upon detecting a defect, the system can dynamically adjust parameters such as applied temperature or placement trajectory to mitigate issues and ensure consistent quality in subsequent layers of composite material.

In essence, this dual robot system represents a paradigm shift in composite manufacturing, combining cutting-edge robotics with advanced sensing technologies to pave the way for a new era of precision and reliability in producing complex composite components. As industries increasingly demand lighter, stronger, and more durable materials, this innovative approach stands ready to meet and exceed those challenges head-on.

6.3.
Rework

Rework of any defects found is a crucial step in the inspection process and takes a skilled technician to accomplish it adequately. Recent works have analyzed the intricacies of defect rework in AFP [6.32]. This experimental study focused on reworking gap and overlap type defects (the most common occurrences during normal AFP operations). It was found that the longer the time since material deposition, the greater the degree of adherence between adjacent tows, making it more challenging to remove a single tow without causing damage. Additionally, peeling off one tow from the top of another without applying additional heat significantly increased the likelihood of negatively impacting the underlying tow.

The study also revealed that tows peeled unevenly or at an angle different from their fiber angle tended to break into separate fibers, which increased the chance of disturbing adjacent tows. When splicing tows around a defective area, extreme care must be taken to avoid damaging the underlying tows by cutting deeper than the depth of one tow. Tows removed from a ply at 90° to the underlying ply tended to induce delamination in the underlying tows.

Once rework is initiated, guidance for the technician is often provided via a laser projection system, as shown in **Figure 6.6**. This system will project items such as tow starts and stops, gaps and overlaps, and ply boundaries. The technician can then use this guidance to perform the necessary rework of any defects that are out of tolerance. It is key that a technician is properly trained for this work. Inaccurate rework of a defect can result in the unintended placement of another defect rather than eliminating the intended one.

Figure 6.6 Laser projection onto a cylindrical tool surface.

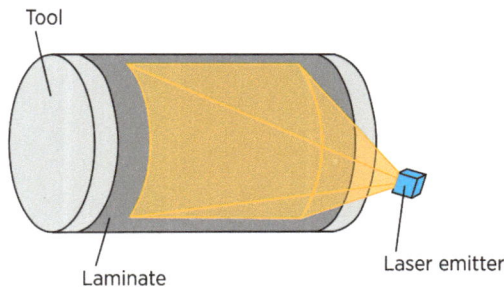

© SAE International.

Some general best practices can be defined based on work done by the authors and results provided by Sacco et al. [6.33], and McArthur et al. [6.32]. In areas of large curvature or high degrees of steering across the tows, only those defects that are the result of improper adhesion to the tool surface or broken or damaged material should be repaired. Any gaps or overlaps that are a result of path planning should not be fixed through rework, but through offline programming. To rework a wrinkle or bridge defect, one should remove the tow entirely and replace it with a new tow that is of the proper length as to accommodate the entire surface of the tool without forcing elongation in the fiber direction. Placement of this tow by hand can be guided by the outline of other bounding tows on the tool surface or potentially a laser projection system. Only large wrinkles should be repaired through reconsolidation or additional heat. The repair of smaller wrinkles is more likely to result in their movement to new sections of the part rather than complete elimination. In some cases, in which there is demand for high-rate production, individual tows can be replaced by rerunning necessary courses with only the tows needing repair being actively placed. However, depending on the complexity and size of the part being manufactured, it may be faster to manually replace a tow rather than the AFP machine. For example, it may be hard for a person to correctly place a steered tow on a fuselage barrel-sized surface, but it would be relatively easy to manually replace a tow on a flat plate. Rerunning the AFP roller over the part can also help with reconsolidating the laminate.

It is recommended to minimize the time between material deposition and reworking to prevent increased difficulty in maintaining separation between adjacent tows. When peeling one tow from the top of another, additional heat should be applied to minimize the negative impact on the underlying tow. Tows should be peeled evenly and in alignment with their fiber angle to avoid breaking into separate fibers and disturbing adjacent tows. When splicing tows around a defective area, extreme caution should be exercised to avoid damaging underlying tows by cutting deeper than the depth of one tow.

6.4.
Utilizing Inspection Data

The utilization of inspection data, along with other key process data, is vital in the effort of continuous improvement of the AFP process. Inspection data can be linked back to design inputs, process parameters, and any number of process inputs. This concept will be expanded on in the last chapter. The data gathered from inspection can be nearly useless unless it is given some context in the form of defect type and location. For example, the authors developed a data mapping technique for projecting profilometry scans back onto a 3D tool surface (see **Figure 6.7**). In this format, it is immediately known the type and location of any defect that was found during inspection. Characteristics of the defects can also be gathered and analyzed to assess whether they are out of tolerance of a given specification. Any defects that were reworked could also be updated and included in the full defect set. Such information would inform the manufacturer of the condition of the delivered part. This also lays the groundwork for future technology insertion such as augmented reality to aid technicians in rework.

Figure 6.7 Aggregate of scan data with defects mapped to a tool surface.

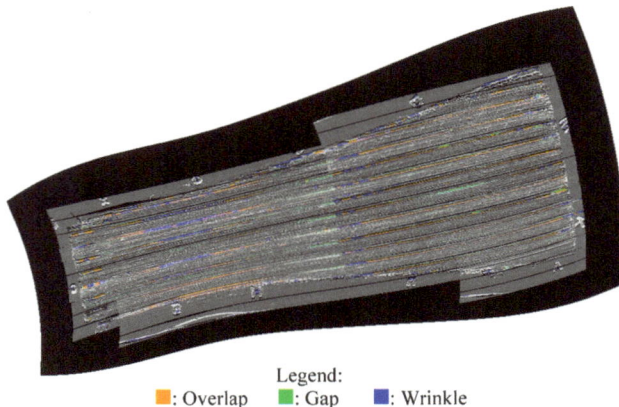

Legend:
■ : Overlap   ■ : Gap   ■ : Wrinkle

Reprinted from Composites Part B: Engineering, 263, Christopher Sacco, Alex Brasington, Max Kirkpatrick, Joshua Halbritter, Matthew Gobold, Ramy Harik, Mapping of multimodalitydata for manufacturing analyses in automated fiber placement, 2023, with permission from Elsevier.

The result of utilizing all aspects of inspection data combined with part and process data is essentially a multilayered array, as shown in **Figure 6.8**, where all data are communicable across the various pillars of AFP. The combination of the models and data also creates not only a virtual process but also a virtual product. All information from design through manufacturing is known and can be queried for future use or provided with the delivery of a part. A manufacturer can then provide a customer with a complete history of the structure they are receiving.

Figure 6.8 Graphical representation of an array of mapped data.

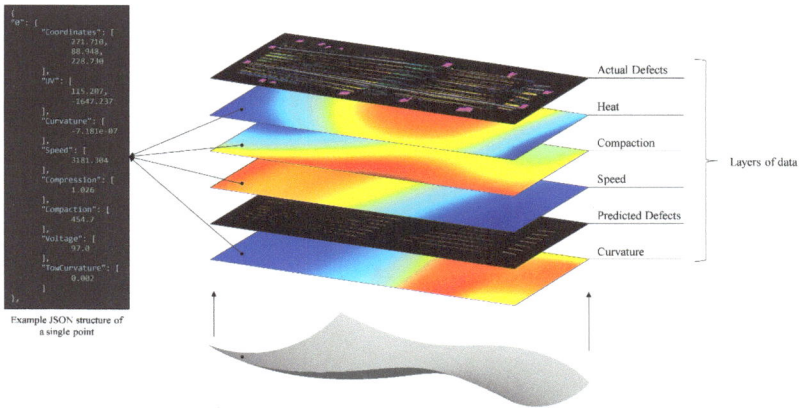

Reprinted from Composites Part B: Engineering, 263, Christopher Sacco, Alex Brasington, Max Kirkpatrick, Joshua Halbritter, Matthew Gobold, Ramy Harik, Mapping of multimodality data for manufacturing analyses in automated fiber placement, 2023, with permission from Elsevier.

6.5.
Other Inspection Methods

AFP-specific inspection is critical for maintaining process and final part quality. However, there are also other techniques that are used for inspecting composite parts after they have been fabricated. These inspection methods are crucial for ensuring part integrity and quality from production to final use. Here, we will provide brief information on typical methods for inspecting composite parts.

Visual inspection is the foundational method in composite manufacturing quality control. It involves direct observation of the composite surface using the naked eye or magnification tools such as magnifying glasses, borescopes, or cameras. This method is essential for detecting surface defects such as cracks, voids, disbonds, delamination, and fiber misalignment. Operators inspect the composite for irregularities that could compromise structural integrity or performance. Visual inspection is typically performed at various stages of manufacturing, from raw material inspection to final product evaluation, ensuring that only components meeting visual quality standards proceed to the next phase of production.

Application

In the previous chapter, we examined how using symmetry in a stacking sequence can help to decouple axial and bending forces. Here, we will dive a bit deeper into this concept with another example.

Assume we have two layups with stacking sequences of $[-45, 90, 45, 0]_T$ and $[-45, 90, 45, 0]_S$, respectively. For practice, you can solve the ABD matrix for both laminates with the following properties: $E_1 = 171.4 \cdot 10^3$ MPa; $E_2 = 9.08 \cdot 10^3$ MPa; $G_{12} = 5.29 \cdot 10^3$ MPa; and $\nu_{12} = 0.32$. The final solutions for the ABD matrices are as follows:

- $[-45, 90, 45, 0]_T$

$$ABD = \begin{bmatrix} 5.71E+8 & 1.78E+8 & 0 & 8.07E+5 & -1.54E+5 & 3.26E+5 \\ 1.78E+8 & 5.71E+8 & 1.49E-8 & -1.54E+5 & -4.98E+5 & 3.26E+5 \\ 0 & 1.49E-8 & 1.97E+8 & 3.26E+5 & 3.26E+5 & -1.54E+5 \\ 8.07E+5 & -1.54E+5 & 3.26E+5 & 4353.09 & 948.44 & -652.82 \\ -1.54E+5 & -4.98E+5 & 3.26E+5 & 948.44 & 1741.8 & -652.82 \\ 3.26E+5 & 3.26E+5 & -1.54E+5 & -652.82 & -652.82 & 1049.5 \end{bmatrix}$$

- $[-45, 90, 45, 0]_S$

$$ABD = \begin{bmatrix} 1.14E+9 & 3.56E+8 & 0 & 0 & 0 & 0 \\ 3.56E+8 & 1.14E+9 & 2.98E-8 & 0 & 0 & 0 \\ 0 & 2.98E-8 & 3.94E+8 & 0 & 0 & 0 \\ 0 & 0 & 0 & 1.41E+4 & 1.01E+4 & -6.53E+3 \\ 0 & 0 & 0 & 1.01E+4 & 2.97E+4 & -6.53E+3 \\ 0 & 0 & 0 & -6.53E+3 & -6.53E+3 & 1.09E+4 \end{bmatrix}$$

With the *ABD* matrices, find the force and moment resultants for each of the two laminates if the midplane stress and strains are given as

$$
\begin{bmatrix}
\varepsilon_x^0 \\
\varepsilon_y^0 \\
\gamma_{xy}^0 \\
\kappa_x^0 \\
\kappa_y^0 \\
\kappa_{xy}^0
\end{bmatrix}
=
\left[
\begin{Bmatrix}
2*10^{-6} \\
-2*10^{-6} \\
6*10^{-7}
\end{Bmatrix} \dfrac{m}{m} \\[2pt]
\begin{Bmatrix}
4*10^{-5} \\
-3*10^{-5} \\
8*10^{-4}
\end{Bmatrix} \dfrac{1}{m}
\right]
$$

Recall that the applied forces and moments are related to the midplane strains and curvatures via the *ABD* matrix:

$$
\begin{bmatrix}
N_x \\
N_y \\
N_{xy} \\
M_x \\
M_y \\
M_{xy}
\end{bmatrix}
=
\begin{bmatrix}
A_{11} & A_{12} & A_{16} & B_{11} & B_{12} & B_{16} \\
A_{12} & A_{22} & A_{26} & B_{12} & B_{22} & B_{26} \\
A_{16} & A_{26} & A_{66} & B_{16} & B_{26} & B_{66} \\
B_{11} & B_{12} & B_{16} & D_{11} & D_{12} & D_{16} \\
B_{12} & B_{22} & B_{26} & D_{12} & D_{22} & D_{26} \\
B_{16} & B_{26} & B_{66} & D_{16} & D_{26} & D_{66}
\end{bmatrix}
\begin{bmatrix}
\varepsilon_x^0 \\
\varepsilon_y^0 \\
\gamma_{xy}^0 \\
\kappa_x^0 \\
\kappa_y^0 \\
\kappa_{xy}^0
\end{bmatrix}
$$

Performing the matrix multiplication results in the final answers as follows:

- $[-45, 90, 45, 0]_T$

$$
\begin{bmatrix}
N_x \\
N_y \\
N_{xy} \\
M_x \\
M_y \\
M_{xy}
\end{bmatrix}
=
\left[
\begin{Bmatrix}
1.08E+3 \\
-5.17E+2 \\
-2.24
\end{Bmatrix} \dfrac{N}{m} \\[2pt]
\begin{Bmatrix}
1.74 \\
0.35 \\
0.74
\end{Bmatrix} N
\right]
$$

- $[-45, 90, 45, 0]_S$

$$
\begin{bmatrix}
N_x \\
N_y \\
N_{xy} \\
M_x \\
M_y \\
M_{xy}
\end{bmatrix}
=
\left[
\begin{Bmatrix}
1.57E+3 \\
-1.57E+3 \\
236.14
\end{Bmatrix} \dfrac{N}{m} \\[2pt]
\begin{Bmatrix}
-4.96 \\
-5.71 \\
8.63
\end{Bmatrix} N
\right]
$$

Now, we can make several observations based on these results. The force resultants N_x, N_y, and N_{xy} are much higher in the symmetric laminate with more plies. This means that the larger laminate can carry significantly higher loads before reaching the same strain and curvature values. The force resultants are higher because the stress is integrated over the thickness of the laminate; therefore, a laminate with more plies can carry higher loads. The moment resultants M_x, M_y, and M_{xy} are also higher in the symmetric laminate with additional lamina. This again can be attributed to the fact that the integration bounds for the D matrix are larger. However, since the second laminate is symmetric, there is no bend-twist and bend-extension coupling. This is beneficial if the goal is for the composite to act more like an isotropic material. The only bending will be a result of the applied moments. Stacking sequence plays a vital role in the performance of the composite.

Ultrasonic testing (UT) is a sophisticated NDT technique used to assess the internal structure of composite materials. It involves transmitting high-frequency sound waves into the material through a transducer. These sound waves travel through the material until they encounter a boundary, such as a defect or the opposite surface. Reflected waves are then detected by the transducer and analyzed to identify internal defects like voids, disbonds between layers, or delamination. UT provides detailed insights into the structural integrity of composites, allowing manufacturers to identify potential defects early in the production process. This method is particularly valuable for its ability to penetrate thick materials and provide precise defect localization and sizing.

Thermography employs IR imaging to detect variations in surface temperature across composite materials. It works on the principle that different materials or defects absorb and emit IR radiation differently, creating thermal patterns that can be visualized using an IR camera. During inspection, thermography identifies defects such as delamination, voids, or foreign inclusions based on their distinct thermal signatures. This method is effective for detecting defects that may not be visible to the naked eye but can affect the composite's structural integrity or thermal performance. Thermography is non-contact and provides real-time imaging, making it suitable for both laboratory and field inspections of composites.

X-ray and computed tomography (CT) are advanced imaging techniques used to obtain detailed cross-sectional images of composite materials. X-ray imaging uses penetrating radiation to create two-dimensional (2D) images of internal structures, while CT scanning combines X-ray data from multiple angles to generate 3D images with higher resolution and depth information. These techniques allow manufacturers to visualize internal defects such as voids, fiber misalignment, or manufacturing anomalies that could compromise the structural integrity of the composite. X-ray and CT scanning are crucial for quality assurance, providing comprehensive insights into the internal structure of complex composite components and ensuring compliance with stringent industry standards.

Shearography is an optical inspection method that utilizes laser light to detect surface deformations in composite materials. During inspection, a laser beam is directed onto the composite surface, and the reflected light interference patterns are analyzed. These interference patterns change with surface deformation, allowing operators to identify defects such as delamination, voids, or impact damage. Shearography provides rapid, real-time assessment capabilities, making it suitable for inspecting large composite structures or components during manufacturing processes. This method is particularly useful for detecting defects that alter the surface structure or mechanical properties of the composite, ensuring product quality and reliability.

AE testing is a NDT method that monitors high-frequency acoustic signals emitted by materials when subjected to stress or loading conditions. During testing, AE sensors detect and record these signals, which are analyzed to identify active defects such as cracks, fiber breakage, or matrix cracking within composite structures. AE serves as an early warning system for potential structural failures, providing valuable insights into the structural health and integrity of composite components. This method is particularly useful for monitoring the progression of defects under operational conditions, ensuring safety and reliability in critical applications.

Tap testing is a simple yet effective method for evaluating the structural integrity of composite materials. It involves tapping or striking the surface of the composite with a handheld instrument and analyzing the resulting sound response. Operators listen for changes in the acoustic signature, which can indicate the presence of defects such as delamination or voids near the surface. Tap testing provides immediate feedback on the integrity of composite components, complementing other inspection methods by offering a quick and practical means of defect detection during production or field inspections.

Moisture content analysis is essential for assessing the presence and level of moisture absorbed by composite materials. Excess moisture can compromise the mechanical properties of composites, leading to issues such as swelling, reduced strength, or delamination over time. Various methods, including moisture meters or gravimetric analysis, are used to measure and monitor moisture levels within composite materials. By controlling and minimizing moisture content, manufacturers can ensure that composites maintain their intended performance characteristics and longevity in diverse environmental conditions.

These inspection methods collectively form a comprehensive toolkit for quality assurance in composite manufacturing. By employing a combination of visual, NDT, and analytical techniques, manufacturers can identify and mitigate defects early in the production process, ensuring that composite materials meet stringent performance standards and deliver optimal reliability in various industrial applications.

AFP Inspection

Duration: 26 minutes

Description: This video, part of "Composites A-Z: 30 Days of Composites," explores AFP inspection methods. It highlights the challenges of manual inspection and introduces automated techniques like thermography, profilometry, and lasers, emphasizing their role in improving defect detection and part quality.

Scan the QR code below to watch the video.

References

6.1. Brüning, J., Denkena, B., Dittrich, M.A., and Hocke, T., "Machine Learning Approach for Optimization of Automated Fiber Placement Processes," *1st CIRP Conference on Composite Materials Parts Manufacturing* 66 (2017): 74-78, doi:https://doi.org/10.1016/j.procir.2017.03.295.

6.2. Schmidt, C., Schultz, C., Weber, P., and Denkena, B., "Evaluation of Eddy Current Testing for Quality Assurance and Process Monitoring of Automated Fiber Placement," *Composites Part B: Engineering* 56 (2014): 109-116, doi:https://doi.org/10.1016/j.compositesb.2013.08.061.

6.3. Rudberg, T., Cemenska, J., and Sherrard, E., "A Process for Delivering Extreme AFP Head Reliability," SAE Technical Paper 2019-01-1349 (2019), doi:https://doi.org/10.4271/2019-01-1349.

6.4. Cemenska, J., Rudberg, T., and Henscheid, M., "Automated In-Process Inspection System for AFP Machines," *SAE Int. J. Aerosp.* 8, no. 2 (2015): 303-309, doi:https://doi.org/10.4271/2015-01-2608.

6.5. Halbritter, A. and Harper, R., "Big Parts Demand Big Changes to the Fiber Placement Status Quo," in *SME Composites Manufacturing*, Mesa, AZ, 2012.

6.6. Rudberg, T., Neilson, J., Henschied, M., Cemenska, J. et al., "Improving AFP Cell Performance," *SAE Int. J. Aerosp.* 7, no. 2 (2014): 317-321, doi:https://doi.org/10.4271/2014-01-2272.

6.7. Hu, E. and Haifeng, F., "Surface Profile Inspection of a Moving Object by Using Dual-Frequency Fourier Transform Profilometry," *Optik* 122, no. 14 (2011): 1245-1248, doi:https://doi.org/10.1016/j.ijleo.2010.08.007.

6.8. Sacco, C., "Machine Learning Methods for Rapid Inspection of Automated Fiber Placement Manufactured Composite Structures," University of South Carolina, 2019, https://scholarcommons.sc.edu/etd.

6.9. Sacco, C., Radwan, A.B., Harik, R., and Van Tooren, M., "Automated Fiber Placement Defects: Automated Inspection and Characterization," in *International SAMPE Technical Conference*, Long Beach, 2018, https://ntrs.nasa.gov/search.jsp?R=20190027133.

6.10. Gardiner, G., "Coriolis Composites and Edixia Develop Inline Inspection for AFP," Composites World, 2020, accessed December 18, 2024, https://www.compositesworld.com/articles/coriolis-composites-and-edixia-develop-inline-inspection-for-afp.

6.11. Electroimpact, "Technology: View ElectroImpact's Catalogue of Advance Composites Manufacturing Products," 2021, accessed December 18, 2024, https://www.electroimpact.com/products/composites-manufacturing/technology.

6.12. Denkena, B., Schmidt, C., Völtzer, K., and Hocke, T., "Thermographic Online Monitoring System for Automated Fiber Placement Processes," *Composites Part B: Engineering* 97 (2016): 239-243. https://doi.org/10.1016/j.compositesb.2016.04.076.

6.13. Gregory, E.D. and Juarez, P.D., "In-Situ Thermography of Automated Fiber Placement Parts," *AIP Conference Proceedings* 1949, no. 1 (2018): 060005, doi:https://doi.org/10.1063/1.5031551.

6.14. Schmidt, C., Denkena, B., Völtzer, K., and Hocke, T., "Thermal Image-Based Monitoring for the Automated Fiber Placement Process," *1st Cirp Conference on Composite Materials Parts Manufacturing* 62 (2017): 27-32, doi:https://doi.org/10.1016/j.procir.2016.06.058.

6.15. Juarez, P.D. and Gregory, E.D., "In Situ Thermal Inspection of Automated Fiber Placement Manufacturing," *AIP Conference Proceedings* 2102 (2019): 120005, doi:https://doi.org/10.1063/1.5099847.

6.16. Maass, D., "Progress in Automated Ply Inspection of AFP Layups," *Reinforced Plastics* 59, no. 5 (2015): 242-245, doi:https://doi.org/10.1016/j.repl.2015.05.002.

6.17. Krizhevsky, A., Sutskever, I., and Hinton, G.E., "ImageNet Classification with Deep Convolutional Neural Networks," *Advances in Neural Information Processing Systems* 25, no. 2 (2012): 1-9, doi:https://doi.org/10.1016/j.protcy.2014.09.007.

6.18. Dung, C.V. and Anh, L.D., "Autonomous Concrete Crack Detection Using Deep Fully Convolutional Neural Network," *Automation in Construction* 99 (2019): 52-58, doi:https://doi.org/10.1016/j.autcon.2018.11.028.

6.19. Meng, M., Chua, Y.J., Wouterson, E., and Ong, C.P.K., "Ultrasonic Signal Classification and Imaging System for Composite Materials via Deep Convolutional Neural Networks," *Neurocomputing* 257 (2017): 128-135, doi:https://doi.org/10.1016/j.neucom.2016.11.066.

6.20. He, K., Zhang, X., Ren, S., and Sun, J., "Identity Mappings in Deep Residual Networks," in Leibe, B., Matas, J., Sebe, N., and Welling, M. (eds), *Computer Vision – ECCV 2016. ECCV 2016*, Lecture Notes in Computer Science (Including Subseries Lecture Notes in Artificial Intelligence and Lecture Notes in Bioinformatics) (Cham: Springer, 2016), doi:https://doi.org/10.1007/978-3-319-46493-0_38.

6.21. Pan, S.J. and Yang, Q., "A Survey on Transfer Learning," *IEEE Transactions on Knowledge and Data Engineering* 22, no. 10 (2010): 1345-1359, doi:https://doi.org/10.1109/TKDE.2009.191.

6.22. Sun, Q., Liu, Y., Chua, T., and Schiele, B., "Meta-Transfer Learning for Few-Shot Learning," in *Proceedings of the IEEE/CVF Conference on Computer Vision and Pattern Recognition (CVPR)*, Long Beach, 2019, 403-412.

6.23. Krawczyk, B., "GPU-Accelerated Extreme Learning Machines for Imbalanced Data Streams with Concept Drift," *Procedia Computer Science* 80 (2016): 1692-1701. https://doi.org/10.1016/j.procs.2016.05.509.

6.24. Liang, S., Yin, S., Liu, L., Luk, W. et al., "FP-BNN: Binarized Neural Network on FPGA," *Neurocomputing* 275 (2018): 1072-1086, doi:https://doi.org/10.1016/j.neucom.2017.09.046.

6.25. Noronha, D.H., Salehpour, B., and Wilton, S.J.E., "LeFlow: Enabling Flexible FPGA High-Level Synthesis of Tensorflow Deep Neural Networks," 2018.

6.26. Pérez, J., Alabdo, A., Pomares, J., García, G.J. et al., "FPGA-Based Visual Control System Using Dynamic Perceptibility," *Robotics and Computer-Integrated Manufacturing* 41 (2016): 13-22, doi:https://doi.org/10.1016/j.rcim.2016.02.005.

6.27. Posewsky, T. and Ziener, D., "Throughput Optimizations for FPGA-Based Deep Neural Network Inference," *Microprocessors and Microsystems* 60 (2018): 151-161, doi:https://doi.org/10.1016/j.micpro.2018.04.004.

6.28. Wienbrandt, L., Kässens, J.C., Hübenthal, M., and Ellinghaus, D., "1000× Faster than PLINK: Combined FPGA and GPU Accelerators for Logistic Regression-Based Detection of Epistasis," *Journal of Computational Science* 30 (2019): 183-193, doi:https://doi.org/10.1016/j.jocs.2018.12.013.

6.29. Sacco, C., Baz Radwan, A., Anderson, A., Harik, R. et al., "Machine Learning in Composites Manufacturing: A Case Study of Automated Fiber Placement Inspection," *Composite Structures* 250 (2020): 112514, doi:https://doi.org/10.1016/j.compstruct.2020.112514.

6.30. Sacco, C., Baz Radwan, A., Beatty, T., and Harik, R., "Machine Learning Based AFP Inspection: A Tool for Characterization and Integration," in *International SAMPE Technical Conference*, Charlotte, May 2019.

6.31. Sacco, C., Radwan, A.B., Anderson, A., Harik, R. et al., "NDE Inspection of AFP Manufactured Cylinders Using and Intelligent Segmentation Algorithm," in *SAMPE 2020*, Virtual Conference, 2020.

6.32. McArthur, S., Mehnen, J., Yokan, C., and Bomphray, I., "An Overview of Current Research in Automated Fibre Placement Defect Rework," *Procedia Computer Science* 232 (2024): 2167-2180, doi:https://doi.org/10.1016/j.procs.2024.02.036.

6.33. Sacco, C., Brasington, A., Saidy, C., Kirkpatrick, M. et al., "On the Effect of Manual Rework in AFP Quality Control for a Doubly-Curved Part," *Composites Part B: Engineering* 227 (2021): 109432, doi:https://doi.org/10.1016/j.compositesb.2021.109432.

AFP Equipment at the University of Concordia in Canada.

Chapter 07

Curing and Testing

7.1.
Curing and Consolidation

This section examines various curing and consolidation processes employed in composite manufacturing, where curing typically refers to thermoset composites and consolidation to thermoplastic composites. Primary curing methods include autoclave curing, utilizing combined pressure and elevated temperature; oven curing, employing heat alone; and infusion, a vacuum-assisted process for dry fiber composites. Detailed explanations of these techniques are provided in the subsequent section. Additionally, in-situ processing, technically a consolidation method for thermoplastic composites, is also briefly discussed here, as it similarly yields the final part geometry.

7.1.1.
Autoclave

Autoclave curing is the most common method in industry for AFP-manufactured parts, and prepreg composite parts in general, due to the increased mechanical properties that are achieved. For example, the fuselage sections of the Boeing 787 are cured in a very large autoclave. The improvement in properties is a result of higher pressures, when compared to vacuum bagging alone, combined with the temperature that significantly reduces voids. The temperature gets the resin to flow better through the material while the pressure assists in getting the resin to flow uniformly through the part. More pressure also provides better laminate compaction against the mold in areas of complex geometry. An example of an autoclave and a schematic of the internal function is shown in **Figure 7.1**. The laminate is vacuum bagged before being put into the autoclave, after which further pressure and temperature are applied. Vacuum bagging is a technique where the laminate is placed into the sealed bag (or mold), and then, the air is removed to create the vacuum. Each material has a manufacturer-recommended cycle for curing in an autoclave, and this cycle can be further optimized for reduced cycle times or to achieve optimal cure for complex parts.

Figure 7.1 Example of (a) an autoclave and (b) internal schematic.

(a)

© SAE International.

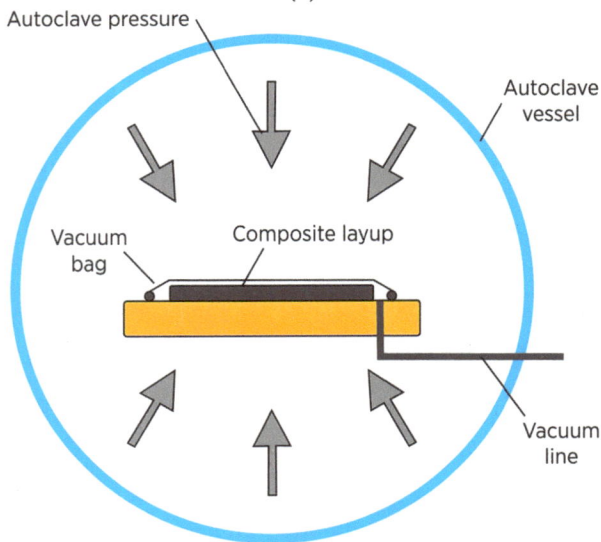

Autoclave pressure

Autoclave vessel

Vacuum bag

Composite layup

Vacuum line

(b)

© SAE International.

The definition of curing cycles, also referred to as recipes, is a critical requirement for using autoclaves. The recipes define the temperature ramp rate, hold temperature, pressure, and time criteria for the autoclave. An example is shown in **Figure 7.2**. Here, the blue line is the autoclave temperature, the yellow line is the vacuum bag pressure, and the gray line is the autoclave pressure. This example is referred to as a one-hold cure cycle as there is only a single holding temperature. An example of a two-hold cure cycle will be shown in the next section. All materials are different, and they require different cycles that have been experimentally developed. Some material needs the initial hold in curing (sometimes referred to as dwell time) to enhance flow and wet-out for uniform distribution, air removal of potentially trapped airs/volatiles, stress relaxation, and certain specific chemical reactions. The recommended cure recipe for each material will be different and is provided by the material supplier. The goal is to heat up the resin to the point that the viscosity has reduced enough for the resin to flow to fill any voids and allow for compaction of the laminate. The temperature is then held to allow for complete curing of the material.

Figure 7.2 Recommended autoclave cure cycle for Toray 3960 prepreg system [7.1]. Temperature is shown in blue, pressure in gray, and vacuum in yellow.

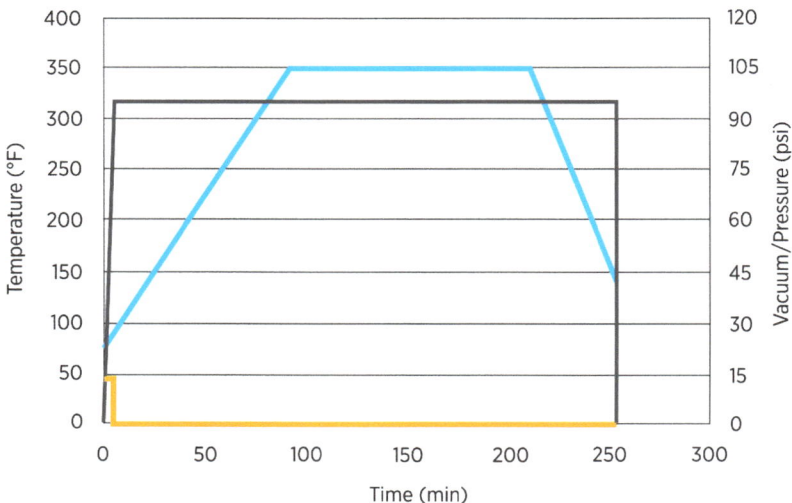

© SAE International.

7.1.2.
Oven

Oven curing is considered an OOA process. In this process, the oven only provides temperature and not additional pressure as seen in an autoclave. Due to the lack of pressure, the mechanical properties are often lower due to the opposite effect of those presented in the autoclave section. However, material manufacturers are developing new material systems for OOA processes that are approaching the properties of their autoclaved counterparts. An example of an oven and an internal schematic is shown in **Figure 7.3**. The use of an oven is typically more cost-effective than an autoclave.

Figure 7.3 Example of (a) an oven and (b) an internal schematic.

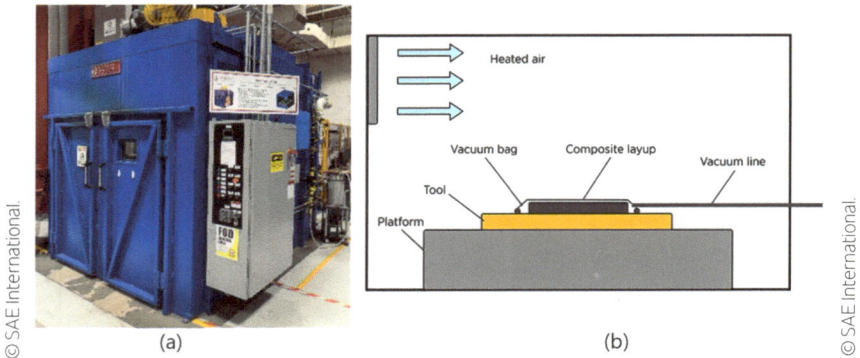

© SAE International.

(a)

(b)

© SAE International.

Again, the cure cycle definition is key to the proper curing of the composite. **Figure 7.4** shows two recommended oven cure cycles for Toray's 2510 OOA prepreg system. The oven temperature is represented by the blue line, while the yellow line indicates the pressure inside the vacuum bag. **Figure 7.4(a)** again shows a one-hold cure cycle, while **Figure 7.4(b)** shows a two-hold cure cycle. Two-hold cycles are recommended for thick laminates to reduce the risk of interlaminar voids and excessive exothermic reactions during cure. Since the material is undergoing a chemical reaction resulting in heat generation, it can often exceed the allowable maximum temperature if not accounted for properly. This point is often referred to as the exotherm temperature during the cure cycle.

Figure 7.4 Example of (a) one hold and (b) two hold cure cycles for Toray 2510 OOA prepreg system [7.2].

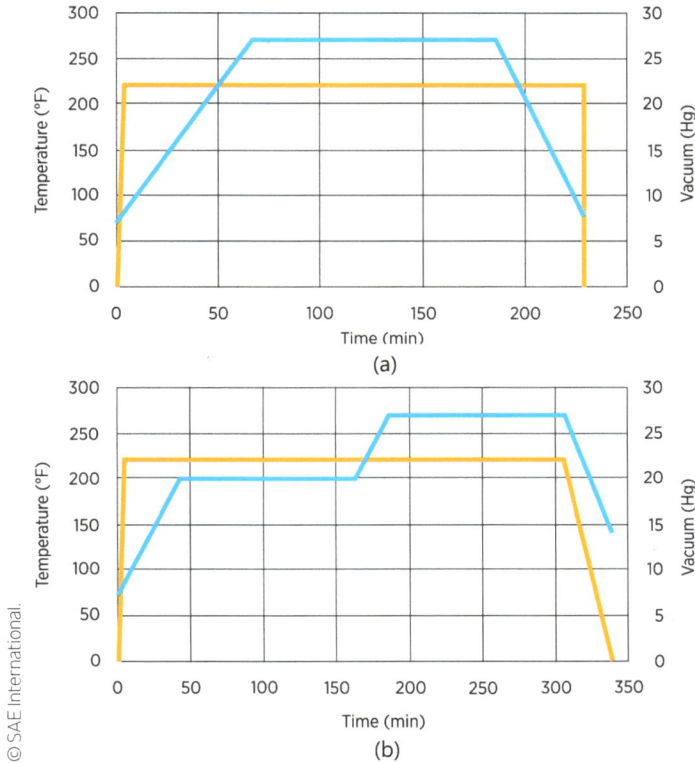

Curing

Duration: 24 minutes

Description: This video, part of "Composites A-Z: 30 Days of Composites," examines curing processes for composite matrices. It covers Resin Transfer Molding (RTM) and Vacuum Assisted RTM (VARTM) for resin infusion into dry fiber, and oven and autoclave methods using prepreg, emphasizing temperature, pressure, and time variables.

Scan the QR code below to watch the video.

7.1.3.
Infusion

As previously detailed, dry fiber materials can be laid up using AFP. Once the AFP is completed, these materials require a resin infusion process to impregnate the fibers with a resin matrix and create the composite structure. While infusion can be—rarely—performed within an autoclave environment, this process is often used as an OOA process. With today's resin infusion processes, manufacturers are no longer constrained by prepreg materials and autoclave size constraints, and can achieve good structural performance. These advancements in recent years are due to both advances in materials and equipment. Once the placement of the dry fiber layup is complete, the fiber bed is bagged similar to the diagram in **Figure 7.5**. Resin is then introduced through an inlet, and the vacuum is pulled from a vacuum port. The resin is then gradually transferred across the laminate. A complete impregnation through the thickness is crucial for the structural integrity of the part. An incomplete impregnation will cause dry spots of unbonded fibers and is one of the most common defects that could form during an infusion process. The placement of resin inlets and vacuum ports plays a large role in the quality of infusion that is achieved in the part. The placement of these ports is typically simulated to determine the optimal number and locations of inlets and vacuum ports. The infusion process can be accomplished at room temperature or can be put into an oven. This process can often be seen in the wind energy industry for infusing large wind turbine blades.

Figure 7.5 Resin infusion bagging.

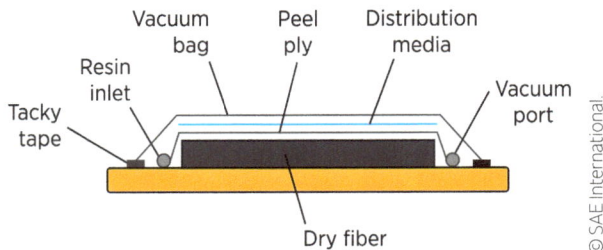

Did You Know?

Audio und werbung/Shutterstock.com.

The energy industry is a significant user of composite materials, especially in wind turbine blades. The scale of wind turbine blades is massive, with General Electric producing glass fiber blades that are over 70 m long. The first step in production is for workers to make the blades from a combination of fiberglass fabric and balsa wood. The use of the materials keeps the blades light but strong. The blade is then covered with an airtight vacuum bag configuration with a network of tubes that will distribute the resin. Appropriate temperature, pressure, and vacuum are applied and monitored while the resin flows throughout the part. Once cured, the blade is ready for final preparations and delivery to harness energy for people around the world.

7.1.4.
In Situ Processing/Consolidation

In situ processing is enabled with the use of thermoplastic materials and is a large focus of current research. Thermoplastic resins allow for repeated melting due to their inherent lack of crosslinking of the polymer structure. Therefore, the heater on the AFP head can heat the resin above its glass transition temperature and place the incoming material. After cooling, the resulting placement has created a laminate that is already in a cured state. This process eliminates the need for an additional curing cycle and the associated hardware, drastically reducing processing times and costs. However, this is a very complicated process, and current technologies have not demonstrated adequate mechanical properties at sufficient processing speeds for adoption in industry. Currently, an additional oven or autoclave processing step is still required, but can be done with shorter process cycle times. It is also worth noting that this process is essentially curing a laminate ply by ply. Depending on the stacking sequence, this can result in locally unbalanced laminates causing warpage during manufacturing.

7.2.
Testing and Certification

The process of testing and certification of a given structure is an in-depth and lengthy process with specific steps being highly dependent on the structure's type and use case. In this section, we will only provide an overview of what is involved in testing and certifying an AFP-manufactured structure. References to supporting documentation will be provided throughout, and the reader is referred to these for detailed information.

Often, a testing campaign utilizes a building block approach defined by the Federal Aviation Administration (FAA) [7.3]. This approach can be represented as a "testing pyramid," as shown in **Figure 7.6**. The large quantity of tests needed to provide a statistical basis comes from the lowest levels, i.e., coupons and elements. Tests performed at these levels evaluate the constituent-, lamina-, and laminate-level properties. Constituent tests include evaluating individual properties of fibers, fiber forms, matrix materials, and fiber–matrix preforms. Key properties, for example, include fiber and matrix density, and fiber tensile strength and tensile modulus. Lamina tests evaluate the properties of the fiber and matrix together in the composite material form. Prepreg properties are often included in this level but are sometimes broken out in a separate level. Key properties here include fiber areal weight, matrix content, void content, cured ply thickness, lamina tensile strengths and moduli, lamina compressive strengths and moduli, and lamina shear strengths, moduli, and Poisson's ration. Next, the laminate-level tests characterize the response of the composite material in a given laminate design. Key properties to be evaluated include tensile strengths and moduli, compressive strengths and moduli, shear strengths and moduli, interlaminar fracture toughness, and fatigue resistance.

Moving up the pyramid transitions the tests into more complex structural elements and structural subcomponents. The evaluation of structural elements assesses the ability of the material to withstand common laminate discontinuities. These properties include open and filled hole tensile strengths, open and filled hole compressive strengths, compression after impact strength, joint bearing, and bearing bypass strengths. Structural subcomponents evaluate the behavior and failure mode of increasingly more complex structural assemblies with the specific tests being highly application specific. These subcomponents then combine to form the final structural components. For example, a component of an aircraft could be a completed full-scale fixed wing or a rotor blade.

Figure 7.6 Schematic diagram of building block tests.

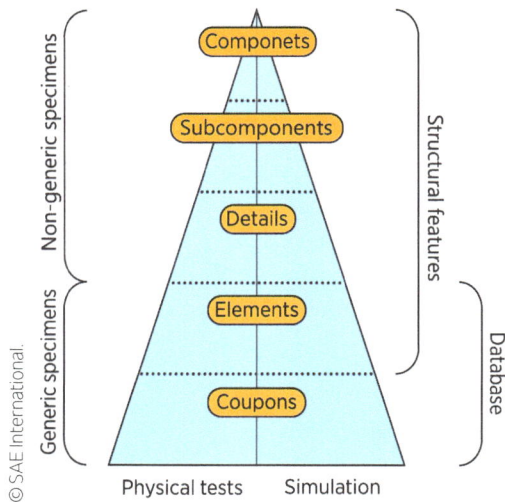

We can further investigate testing and certification by looking at a building block test based on the one presented in [7.4] with the associated FAA documentation [7.5]. **Figure 7.6** is generic, and **Figure 7.7** provides an example of a building block program for an aircraft from initial material selection through final certification.

Figure 7.7 Example of building block test program for an aircraft.

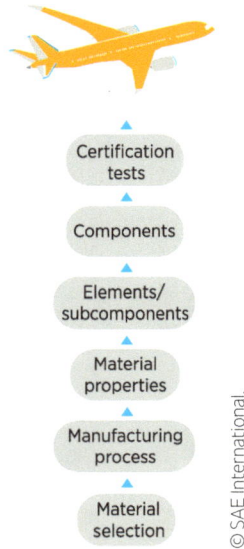

Certification tests

Components

Elements/ subcomponents

Material properties

Manufacturing process

Material selection

© SAE International.

The first step is to select the combination of material and manufacturing process that will be used to fabricate the structure. It is key to look at both items as a package since they will intricately combine during manufacturing. Selection must also consider an evaluation of part requirements and cost goals. Part requirements can consist of loads that must be achieved, structural size, and final part quality. Cost goals are typically associated with upfront capital cost, manufacturing cost per structure (wing skin, fuselage barrel, etc.), and overall cost of the complete product (an aircraft in this case). Certain material and process selections, for instance, AFP, will greatly affect the efficiency and quality of the resulting product, directly correlating to product cost. Further, material and process maturity are a factor. Material allowables for design as well as process reproducibility will be required. If a material or process lacks maturity, material allowables may not be available or the immature process can be unreliable in terms of final part quality. Experience from previous programs can also be used if mature materials and processes are chosen. The outcome of this first step will be candidate materials and processes of which a final selection must be made based on program requirements and budget, available expertise, and structural requirements. Throughout the rest of this example, we will assume that the AFP process has been chosen along with a compatible material system.

Figure 7.8 ElectroImpact's Kinston-5.

Courtesy of ElectroImpact.

Next, the manufacturing process must be developed, also referred to as process development. For AFP, one of the more critical steps of process development is defining the toolpaths and process limits. As mentioned earlier, the definition of toolpaths for a structure is one of the most time-consuming aspects of AFP and must be done with the utmost care. The process limits correspond to upper and lower values for process parameters discussed in an earlier chapter. Mainly, parameters associated with applied temperature, compaction pressure, and layup speed must be established. Recall that these parameters are interconnected and must be chosen as a set and not as individual process inputs. The feasibility of the parameters across the chosen structural geometry must also be considered. Once chosen, mechanical properties must be developed at the upper and lower limits. This is typically accomplished through manufacturing test panels that can then be utilized to perform mechanical testing for desired properties.

Often, composite materials must receive limited "requalification" every time a process parameter is altered, at least to show equivalency. Due to the statistical variability inherent in composites, even relatively small changes during manufacturing can alter the load path or failure mode and produce statistically significant changes in composite properties. A demonstration of reproducibility within the chosen limits is also required. This is accomplished through sub-scale or full-scale builds of the chosen structure. At this point, critical steps, tools, and equipment are defined. Steps include detailed process plans that guide technicians through the complete manufacturing process. Lastly, necessary inspection and quality control tools and processes must be developed. A large portion of AFP inspection is still accomplished visually. However, automated processes are ever-increasing and must be defined at this stage. This includes any equipment, sensors, and planning for proper quality control of the final part. Once this stage is complete, a process specification will be defined. For most companies, this will define the process and rules that must be followed throughout the life of a process.

It is also required to develop a database of material properties for the chosen system if it does not already exist. This development often happens in tandem with the development of the manufacturing process. Firstly, physical and chemical properties must be determined. For AFP material systems, this includes width, density, viscosity, cure kinetics, out time allowance, material tack, and glass transition temperature. Each of these properties will affect the process parameter ranges and process plans developed for the process. This shows the necessity, especially for AFP, that a material system and manufacturing process be defined concurrently. Environmental sensitivity is also key. Properties such as moisture resistance, upper use temperatures, and humidity affects will define the necessary conditions required for the AFP manufacturing environment.

Typically, AFP manufacturing is performed in a controlled environment where temperature and humidity can be readily controlled.
The mechanical properties of the material, including strength, modulus, damage tolerance, and fatigue resistance, must be defined for the material when it is used with AFP. The exact manufacturing process can have positive and negative effects on nominal material behavior. This is especially true when comparing broad-good materials with the slit materials used in AFP. When defining these properties, one must examine all critical modes and environments that the structure will be subject to. For instance, material properties in a hot and wet environment will differ from those in a cold and dry environment. All of these data will then be utilized to develop design requirements that will guide the design engineers when developing the structure. When defining allowables for AFP, it is also critical to consider the effects of defects. If common defects such as gaps and overlaps are not correctly incorporated, the allowables may not properly reflect the actual performance of the structure. The outcome of this step is a complete material specification and a set of design allowables. Before continuing to the next step in the build block certification program, one must also define specifications for the repair of the material used within a structure. This should include a defect and damage sensitivity study to define the tolerances allowed before a structure must be repaired. Procedures and specifications should also be developed for specific material systems.

At this point, the material system and manufacturing process have been fully developed, and specifications exist for both that will be implemented during the manufacturing of parts. Firstly, elements and subcomponents must be fabricated with the chosen material and process. These elements will provide validation of the analyses done during the development of the structure. This includes validating if the as-manufactured components align with the as-designed and have adequate mechanical properties. Such an effort will also refine the material, process, and quality specifications previously developed. Further, these elements and subcomponents act as a reduction in risk for manufacturing and assembly prior to the ramp up to full-scale production.

The next task in the testing and certification program is the fabrication of actual components. At this point, there will be a demonstration of repairs, component-level mechanical performance, system interfaces, damage tolerance, and further validation of analyses. These components will typically undergo a destructive evaluation to intricately study the structure and ensure it is ready for certification. The effort of fabricating actual components further reduces risk and refines processes to their final form.

With the material and processes now in their final production form, it is time to certify the structures that are being fabricated. To pass the certification, the structures must adequately perform when subject to their load cases. This includes static, dynamic, and fatigue loads. If any of these tests fail, the structure will not be qualified and may need to undergo redesign, or the manufacturing process may need to be altered to improve quality. Assuming all structures pass certification tests, in this example the aircraft will go on to flight testing. At this point, flight clearance will be given based on the full package of data gathered during the entirety of the building block test and certification program.

Infusion

Duration: 21 minutes

Description: This video, part of "Composites A-Z: 30 Days of Composites," covers infusion, a closed molding process injecting resin into dry fiber. Methods include Vacuum Infusion (low rate), Resin Transfer Molding (mid rate), and Reaction Injection Molding (high rate). VARTM relies on permeability, viscosity, and pressure.

Scan the QR code below to watch the video.

7.3.
Conclusion

As a final point in this chapter on the pyramid process for AFP and composite laminates, it is worth mentioning the leading role of the FAA in aerospace certification. The FAA's demanding requirements on composite layups, particularly for new or generic configurations, can lead to lengthy and potentially uncertain certification paths. Deviations from widely recognized non-rosette ply layups can necessitate a significantly larger volume of testing and scrutiny than might otherwise be anticipated. This is one of the primary reasons for aerospace practice to standardly favor the use of quasi-isotropic laminates. These balanced laminates, with their more predictable and better understood mechanical response in all of the in-plane directions, will tend to simplify certification relative to more exotic stacking sequences. Therefore, in canvassing the design space of composite structures, engineers ought to carefully consider the FAA certification environment and the practical advantages of industry-standard quasi-isotropic laminates in order to attain a simpler and more efficient path to airworthiness.

References

7.1. TORAY, "3960 Prepreg System Data Sheet," Toray Composite Materials America, Inc., Tacoma, WA, 2022.

7.2. TORAY, "2510 Prepreg System Data Sheet," Toray Composite Materials America, Inc., Tacoma, WA, 2020.

7.3. Federal Aviation Administration, "AC 20-107B - Composite Aircraft Structure," Washington, DC, 2009.

7.4. Joyce, P., "Composite Material Qualification Process," United State Naval Academy, Annapolis, 2003, Retrieved from https://www.usna.edu/Users/mecheng/pjoyce/composites/Short_Course_2003/X_PAX%20Short%20Course_Material%20Qualification.pdf.

7.5. Federal Aviation Administration, "AC 23-20 - Acceptance Guidance on Material Procurement and Process Specifications for Polymer Matrix Composite Systems," Washington, DC, 2019.

Clemson Composites Center in Greenville, South Carolina.

Chapter 08

The Future of AFP

Thinking about the future of composite manufacturing and integration of automation and robotics, this chapter represents the foundational digital workflow and data architecture needed to move composite manufacturing, specifically AFP, toward a new paradigm.

Successful adoption of AFP into new and innovative domains requires understanding the steps needed to mathematically convert the defined specifications (process parameters, toolpaths, etc.) into machine trajectories. While this is achieved by carefully integrating the previously detailed concepts from process planning and layup strategies, our target in this chapter is to offer an insight into how these different elements/tools actively participate in manufacturing the desired structures.

First, design is continuous in representation, yet discrete in terms of orientation and angle assignments at different locations. In contrast, numerical controllers will ask the machine to generate a trajectory that is mathematically continuous, as the machine will move between positions without "jumps" that might exist in the digital world. In the physical world, while going from one location to another, there is no sampling: It is continuous motion.

This chapter will first introduce the numerical cycle and the different trades needed to communicate, interact, and agree (handshake) to ensure a smooth transition from inception to manufacturing and beyond (recycling). Next, we will detail the AFP Factory of the future.

8.1.
Understanding the Numerical Cycle

Engineering is typically a response to a specified need. This need is often specified in a set of directives that constitutes the starting point of an engineering project. Examples of needs are the requirement to move between two locations (transportation/vehicle), the requirement to control environmental conditions within a facility (heating/ventilation), and the requirement to have a shelter/dwelling (construction). All those needs are defined in a project specification document that defines the functional requirements, as well as the aesthetic ones. A luxury car mixes both functional requirements (the ability to move between two locations) and aesthetic ones (the design should be impressive). Once a set of specifications is defined, the detailed design process initiates. The design process includes the geometrical shapes, as well as material selections, kinematics/motion, and expected behavior of the final structure/product. Geometry is often described/represented in CAD files that are structured with relationships (parent/child). Following the design stage, the process planning phase takes place. Process planning, as described earlier, is the matchmaking between the design and the manufacturing process. This is where manufacturing resources are connected with the design at hand to produce the desired structure. Manufacturing an initial prototype/demonstrator is what ensues the process planning stage. This is typically referred to as prototyping, where all the different elements from design, process planning, and manufacturing are validated. Prototyping is an important validation step and often uses lower-cost manufacturing techniques than the ones that will eventually be used in production. An example would be using hand layup to come up with a structure prior to the purchase of an AFP machine.

Prior to the release of the part from prototyping to the production phase, there are typically alterations and modifications sent back to the design stage, in a circular fashion. Until the prototyping stage releases a product that meets all the requirements as defined in the specifications, the iterative cycle of design, process planning, and prototyping continues. At the attainment of a prototype that meets the requirements, manufacturing now ramps up to meet the production requirements. Often, inspection and quality control might require modifications and changes, and as such, there will be new releases of the product/structure. Additional modifications to the manufacturing plan may also be needed to meet a necessary manufacturing rate, in the case of a high-volume product.

DFM aims to reduce the number and duration of the cycles described above by integrating sufficient insights from manufacturing early in the process, preemptively accounting for resources and restrictions before they manifest themselves in product manufacturing. Limiting the design space based on expected resources might hinder innovation at large; however, it will ensure a shortened release cycle that can be beneficial/viable.

AFP offers precise and efficient manufacturing of composite parts. However, the inherent constraints of the process, such as the available fiber orientations and layup patterns, can sometimes limit the design possibilities. Post-AFP processes can also suppress some of the potential issues that might arise. An example would be the reduction of a pucker in the autoclave process.

ACMA Composites Certification

Duration: 47 minutes

Description: This video, part of "Composites A-Z: 30 Days of Composites," explores the American Composites Manufacturers Association (ACMA) and composite certifications. Andrew Pokelwaldt discusses certification versus compliance, material qualifications, market-specific standards (e.g., aerospace, automotive), and ACMA's role in guidelines and certification.

Scan the QR code below to watch the video.

8.2.
AFP Factory of the Future

With the advent of Industry 4.0, many manufacturing sectors have begun the exploration into the use of smart and digital manufacturing with implementation of machine learning. However, AFP manufacturing, and the remaining composite manufacturing sector, has yet to explore the philosophies of the future of manufacturing. Rather, siloed efforts have been enacted to advance each of the technical challenges associated with AFP, resulting in an open-loop system that is difficult to understand and optimize on a global level.

"Factory of the Future" is an evolving concept that can have various definitions and even names. In this case, it will be defined as a vision for how manufacturers can enhance a production system by making improvements in infrastructure, digitization, and processes to increase efficiency, reduce defects, and optimize the process. This concept materializes within AFP as a closed-loop process in which each pillar of AFP is interconnected through a data pipeline incorporating predictive models, manufacturing data and events, and part quality metrics. This process is depicted through a flowchart in **Figure 8.1**.

Figure 8.1 Graphical representation of closed-loop workflow [8.1].

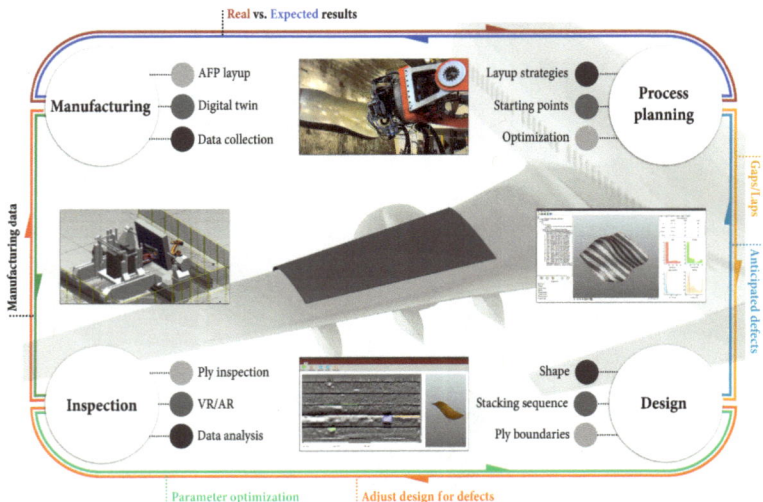

R. Harik, "neXt Automated Fiber Placement: Advancing Composites Manufacturing Towards a New Paradigm", in SAMPE Journal 56 (6) - 6-14. Reprinted by permission from the Society for the Advancement of Material and Process Engineering (SAMPE).

Prior to this framework, design was often the starting point for AFP manufacturing. It is now a single point within the continuous loop of data flow over the manufacturing life of not just one part, but a multitude of parts that can also encompass numerous AFP systems. For example, the data flow in **Figure 8.1** could be a single instance within a large pool of instances with each one adding its own data and analyses to the wealth of data and knowledge. Each iteration of the process then uses this wealth of information to make informed decisions within each pillar.

After an initial design (which includes stress analysis), the communication between design and process planning initiates to generate the most optimal tool paths to be manufactured via the AFP machine. These toolpaths are translated simultaneously with expected empirical results that are to be the outcome of the physical process. This task of matchmaking the designed part with acceptable manufacturing inputs is often manual and left up to the designer and process planner, leading to a time-consuming back and forth to achieve an optimal result. Autonomous optimization and data sharing at this phase can eliminate the highly prevalent habit of designers failing to account for manufacturing limitations at early stages of design. Such an optimization can also streamline getting the part from the design board to the manufacturing floor. Doing so can drastically reduce the development time of a structure along with reducing or eliminating possible tribulations encountered in previous manufacturing trials of other structures. Note that all of this is possible before even encountering the AFP cell.

The next stage of the workflow connects process planning and manufacturing. At this stage of the AFP process, design and process planning have had sufficient back and forth to generate an expected optimal manufacturing plan. Manufacturing is performed while collecting large amounts of data to be communicated to the machine's digital twin or digital shadow such as the one shown in **Figure 8.2**. The main concept of utilizing a digital twin is that the machine can draw the operation data and perform data analysis to propose changes in the manufacturing process. This will require major integration of new automation hardware that can rapidly react to a complex domain with ever-changing material properties and requirements. The current state of the art, combined with safety and certification requirements, does not allow for "on-the-fly" in-process changes to the defined manufacturing procedure. Rather, a digital shadow can be used to accumulate all the manufacturing data to be used in post-process decisions. This allows for immediate deployment of the presented data flow without violations of existing practices and standards.

Figure 8.2 Digital twin in context.

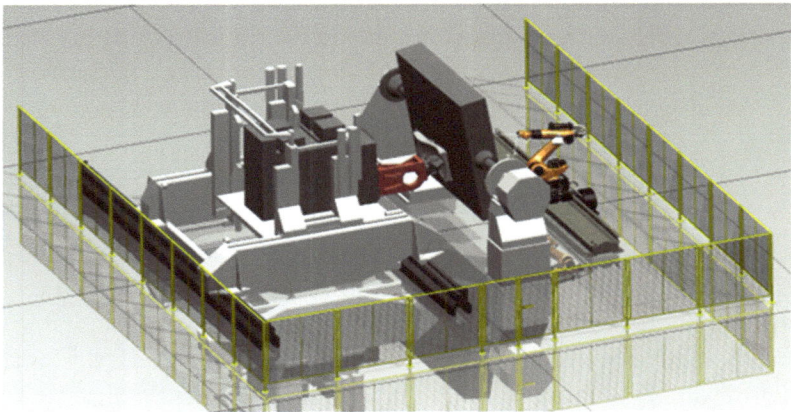

M. Kirkpatrick, A. Brasington, A. Anderson and R. Harik, "Creation of a Digital Twin for Automated Fiber Placement," in CAMX Conference & Exhibition, Orlando, Florida, 21-24 September 2020. Reprinted by permission from the Society for the Advancement of Material and Process Engineering (SAMPE).

With manufacturing now underway, the next stage of integration is connecting manufacturing with information gained from inspection. In this stage, connections between data from the layup of the laminate and the resulting manufacturing defects are correlated. This builds on the continuous effort to create a data-driven composite manufacturing approach. Such an approach would have predictive capabilities that are not possible with current modeling techniques since as-manufactured results would be incorporated. The correlations made at this stage influence the process parameters selected in later manufacturing instances. Also, the functionalities that can be developed within this stage include the representation of data into a virtual or augmented reality world that assists personnel with inspection tasks and understanding system behavior.

Did You Know?

The Office of the Under Secretary of Defense for Research and Engineering released Department of Defense (DoD) instruction 5000.97 effective as of December 2023 that cancels directive 5000.59 on digital engineering.

The directive defines the digital engineering ecosystem. This includes the infrastructure and architecture necessary to support automated approaches for system development, design, testing, evaluation, production, operation, training, and sustainment through the full PLM process. Some of the major elements are the infrastructure elements such as hardware, software, networks, tools, and workforce.

One of the most consequential is the definition of the digital twin part of the digital models section. The definition is as follows: *A digital twin is a virtual representation of a product, system, or process that uses the best available models, sensor information, data collected from the physical system, and input data to mirror and predict system activities and performance over the life of its corresponding physical twin and inform system design changes over time. There can be multiple digital twins of a system, but all digital twins should be based on authoritative sources of information and have clearly defined uses and scopes. Digital twins may vary in fidelity, based on the use case.*

This actually makes one of the strongest and most compelling definitions as it states that digital twins are not merely digital replicas (or shadows); they are dynamic, two-way bridges between the physical and digital worlds. By exchanging data and insights, twins can predict and inform changes in their physical counterparts.

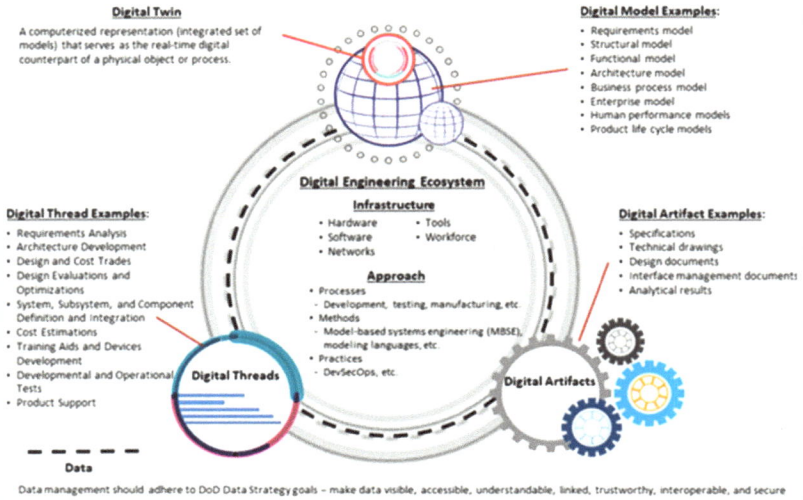

Digital Twin
A computerized representation (integrated set of models) that serves as the real-time digital counterpart of a physical object or process.

Digital Model Examples:
- Requirements model
- Structural model
- Functional model
- Architecture model
- Business process model
- Enterprise model
- Human performance models
- Product life cycle models

Digital Engineering Ecosystem

Infrastructure
- Hardware
- Software
- Networks
- Tools
- Workforce

Approach
- Processes
  - Development, testing, manufacturing, etc.
- Methods
  - Model-based systems engineering (MBSE), modeling languages, etc.
- Practices
  - DevSecOps, etc.

Digital Thread Examples:
- Requirements Analysis
- Architecture Development
- Design and Cost Trades
- Design Evaluations and Optimizations
- System, Subsystem, and Component Definition and Integration
- Cost Estimations
- Training Aids and Devices Development
- Developmental and Operational Tests
- Product Support

Digital Artifact Examples:
- Specifications
- Technical drawings
- Design documents
- Interface management documents
- Analytical results

Digital Threads

Digital Artifacts

Data
Data management should adhere to DoD Data Strategy goals – make data visible, accessible, understandable, linked, trustworthy, interoperable, and secure

Reprinted from DoD Instruction 5000.97.

At this juncture, the integration of data-based approaches will not only help us to obtain new structures with advanced composites, but it will also break many of the barriers hindering the adoption of composites due to their complexity. A hybrid model will require the following:

- Smart hardware that is capable of capturing data on the fly for processing. Think of heating as one of the most complicated processes in AFP. The amount of heat applied makes or breaks the quality of the part in all material systems (thermosets and thermoplastics). Having a smart heating capability that is closed loop is fundamental to achieving tight control of the process [8.2]. Today, the vast majority of heating systems are open loop without tracking.

- Smart software that is capable of interpreting the data on the fly to propose change. We have implemented several algorithms that are based on ML. Such techniques are only as good as the amount of data that are captured, and very often, they are limited. Having a hybrid method can also unlock "data communication" between facilities to enhance mutual productivity.

When you ask data scientists how much data are needed to enhance the model accuracy, you often hear: "It's never enough." Actually, we concur with this assessment, but not for the same reasons as the data scientists. The composite manufacturing lifecycle (design, process planning, manufacturing, and inspection) is so complex that data will never be enough to make any accurate and tangible prediction. This is why working toward a hybrid model, which has the physics complemented with data-driven models, can help unlock the next level of automation for future factories.

Finally, the loop is closed with the connection of inspection and design as described earlier. The data from the inspection are correlated with the design resulting in the optimization of the design parameters. All the data gathered are now available to influence future decisions, with the desire for semi-autonomous decision-making. With the presented connections, a closed-loop AFP workflow is achieved.

Composites with Ingersoll

Duration: 59 minutes

Description: This video, part of "Composites A-Z: 30 Days of Composites," features Ingersoll Machine Tools' AFP technology, focusing on their high-deposition-rate systems designed for large composite structures. It showcases their multi-gantry platforms, precision motion control, and scalable solutions. Ingersoll emphasizes flexibility, speed, and innovation in AFP to meet demanding aerospace and defense manufacturing requirements.

Scan the QR code below to watch the video.

8.3.
Factory of the Future

To drive radical transformation of any industry, manufacturing needs to securely expand beyond the physical boundary of the facility and also benefit from advances in computing and AI. By defining the methods and techniques to extract, augment, and have data transform into knowledge, we can overcome the shortcomings of process-based control.

A factory of the future will need a new digital workflow and a new cyber-infrastructure enabling physics data-based models. The University of South Carolina has been diligently working to create a future factory cell that mimics how rockets should be manufactured. The cell integrates robotics, drones, tracking vision cameras, thermal cameras, augmented sensing capacity in the redesign of the control system, and robot grippers. All this hardware streams data to a cyber-platform that ensures semantics between the collected data. **Figure 8.3** shows the Future Factories Laboratory at the McNAIR Center. A drone is performing additional data collection/inspection on the composite part, as a simulation of how the final platform should operate. Behind the scenes, a cyber-physical-enabled control network administers both the physical cell and the digital twin to synchronize process signals and intelligently actuate field devices by system smart layers (**Figure 8.4**). The cyber-physical system is currently being built to validate several of the proposed concepts. An initial assembly exercise on rockets is seen in the background of **Figure 8.3**.

Figure 8.3 The Future Factories Laboratory at the McNAIR Center of the University of South Carolina.

Figure 8.4 (a) Digital twin of the Future Factories cell, and (b) the actual manufacturing cell.

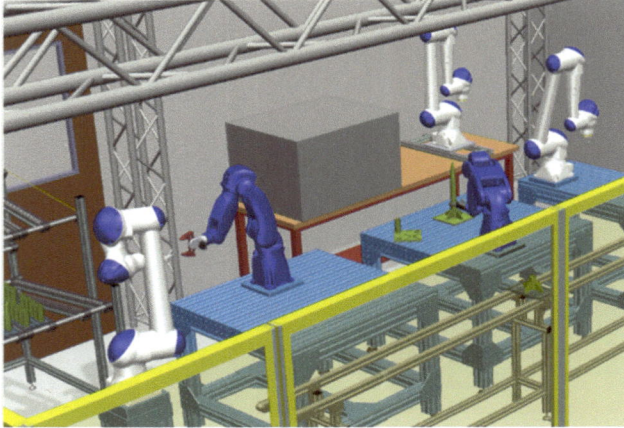

(a)

(b)

R. Harik, "neXt Automated Fiber Placement: Advancing Composites Manufacturing Towards a New Paradigm", in SAMPE Journal 56 (6) - 6-14. Reprinted by permission from the Society for the Advancement of Material and Process Engineering (SAMPE).

8.4
Towards AFP+

This section is taken from the forthcoming article by Harik and Godbold (Manuscript accepted for Composites Handbook 2026) [8.1], "Towards Smart Automated Fiber Placement." The section addresses both the present situation and future directions of AFP intelligence. Whereas acknowledging current challenges, the review, based on statistics from research papers and OEM websites, points out that certain stated capabilities may be marketing claims in the absence of independent verification.

8.4.1
Current AFP Intelligence

Intelligence in AFP systems represents the integration of advanced technologies such as real-time process monitoring, adaptive controls, and data-driven optimization to enhance manufacturing precision and efficiency. This section highlights the current state of AFP intelligence, focusing on smart features and innovations that distinguish modern AFP systems from traditional process capabilities. It should be noted that the information discussed in this section is derived from publicly available sources and marketing materials and does not necessarily stem from independent scientific studies or validated research outcomes.

8.4.1.1
Digital Twin Integration

Digital twin technology has become a cornerstone of intelligent AFP systems, enabling enhanced process simulation, optimization, and decision-making. MTorres employs kinematic digital twins for sequence optimization and collision detection, automation-focused digital twins for streamlined change management and virtual training, and product-specific digital twins that capture manufacturing process data within their Torresfactory platform [8.4]. Similarly, Broetje-Automation's SOUL RMOS Digital Twin supports online path creation, collision detection, and seamless integration with the Siemens Virtual NC-Kernel for accurate virtual machine simulations [8.5]. These systems provide real-time synchronization between physical machines and their virtual counterparts, improving decision-making during concurrent engineering, reducing product design cycle times, and enabling advanced quality analysis and operator training without interrupting production.

8.4.1.2
Real-Time Monitoring and Predictive Maintenance

Real-time monitoring and predictive maintenance are central to intelligent AFP systems. MTorres's Edge Data Analytics platform features self-diagnosis routines that gather machine data, calculate key process indicators, and trigger predictive alarms to enhance maintenance and process reliability [8.6]. This data is seamlessly integrated with customers' MES and ERP systems, enabling efficient decision-making and connectivity. Coriolis Composites' CoDa platform complements these efforts by offering high-frequency data collection and tools like CoDa View for offline analysis and CoDa Export for traceability [8.7]. These systems empower manufacturers to monitor operations in real time, predict equipment failures, and optimize maintenance schedules.

8.4.1.3
Advanced Path Planning and Simulation

Sophisticated path-planning tools play a vital role in intelligent AFP systems, allowing for optimized layup paths and defect prediction. Coriolis Composites' CADFiber and CATFiber solutions enable designers to incorporate manufacturing constraints early in the design process, with support for ply definition imports, advanced analyses, and machine simulations for collision detection and safety validation [8.8, 8.9]. Similarly, Addcomposites' AddPath software offers capabilities for AFP planning, layup visualization, material and cost calculations, and custom robot program generation [8.10]. By integrating Finite Element Analysis (FEA), AddPath optimizes fiber placement, performs stress analysis, and evaluates material performance while enabling online defect detection and predictive maintenance. These tools ensure efficient resource utilization and help mitigate issues such as gaps, overlaps, and angular deviations.

8.4.1.4
Smart Human-Machine Interfaces
Modern AFP systems leverage advanced Human-Machine Interfaces (HMIs) to enhance operator interaction and process control. Coriolis Composites' HMI system provides comprehensive control and traceability, including a three-phase setup with mobile and fixed consoles for real-time monitoring and offline management tools for production and maintenance [8.11]. Emerging alternatives are reshaping traditional HMI setups by incorporating greater flexibility and usability. MTorres's use of Augmented Reality (AR) enables immersive visualization of assembly instructions, part inspections, and rework tasks. These innovations exemplify the industry's shift toward adaptable, operator-friendly interfaces that enhance productivity and user experience [8.4].

8.4.1.5
Data-Driven Quality Assurance
Data-driven quality assurance has become a critical component of AFP intelligence, ensuring traceability and high-quality composite parts with minimal manual intervention. Electroimpact's AFP4.0 philosophy incorporates the RIPITx system, which conducts in-process inspections of defects such as laps, gaps, and foreign object debris, enabling uninterrupted layup and automated ply buy-off [8.12]. Ingersoll Machine Tools' Automated Composite Structure Inspection System (ACSIS) system detects defects with over 99% accuracy and integrates operator feedback through tablets and augmented reality tools to refine its machine learning models [8.13]. Fives' Composite Optical Automated Surface Tracking (COAST) system uses Optical Coherence Tomography (OCT) for real-time, high-quality inspection across a range of composite materials and sensor angles [8.14]. These solutions showcase the integration of advanced inspection and AI technologies, enabling higher efficiency, better defect detection, and continuous process improvement.

8.4.1.6
Challenges and Limitations in AFP Intelligence
Despite significant advancements, current AFP intelligence still faces limitations and gaps that hinder its full potential. One key challenge lies in the lack of standardization across manufacturers in

implementing intelligent systems such as digital twins, real-time monitoring platforms, and advanced path-planning tools. This variability can create barriers to interoperability and limit the seamless integration of AFP systems into broader manufacturing ecosystems.

Additionally, while digital twins and real-time monitoring offer powerful capabilities, their application is often constrained by the complexity of modeling highly intricate geometries or capturing the nuances of specific material behaviors. For example, automated defect detection systems, though highly accurate, may struggle with detecting or resolving issues on complex or irregular part geometries, requiring supplemental manual intervention.

Another critical gap is the reliance on operator feedback and human decision-making during certain inspection and quality assurance processes. While tools like Ingersoll's ACSIS™ and Electroimpact's RIPITx incorporate interactive elements to refine their models, this reliance highlights the current limitations of fully autonomous defect detection and process optimization.

Furthermore, the integration of emerging technologies such as augmented reality (AR) in HMIs and AI-driven predictive maintenance remains in early stages, with varying levels of maturity and adoption across the industry. Broader deployment and refinement of these technologies are necessary to maximize their impact.

Lastly, while AFP intelligence has significantly improved defect detection and prevention, it has not yet achieved a level of adaptability to dynamically adjust layup strategies in real-time without pre-programmed parameters or external interventions. Addressing these limitations will require ongoing research, increased collaboration among manufacturers, and continued investment in advanced algorithms, data analytics, and system integration. Some of the fundamental concepts that can be executed are detailed in the following section.

8.4.2
Proposing a Smart AFP System: AFP+
Integrating intelligence into AFP leverages the "smart" definition established earlier in this chapter to address current AFP challenges.

This involves utilizing advancements in artificial intelligence and machine learning and adopting digital twinning frameworks. The terms used by industry often differ in meaning and application from their intended definitions. This creates inconsistencies within the industry itself and misalignment with the broader research community. As a result, some proposals, despite marketing claims, fail to deliver the necessary technical functionality. To address this, a crucial first step would be to establish a common understanding and consistent use of terminology across all stakeholders. As such, adopting the definitions as stated in DOD 5000.97 would constitute an excellent leveling ground.

AFP, predominantly used in aerospace, is a complex process further complicated by new materials and structures. To expand into new markets, this section identifies key challenges for initial AFP development. While multiple innovative topics such as "toolless AFP" and others remain a long-term goal (potentially relevant for future space manufacturing), this section focuses on realistic advancements achievable within the next decade. The current state of AFP composites manufacturing in the industry is far from ideal. The reality is that AFP remains cost-prohibitive, even with its adoption in certified aircraft like the A350 and B787. The process is hindered by its own flexibility, burdened by an increasing number of digital constraints that limit its true potential. While national programs exist to explore advancements like next-generation resin systems and thermoplastic innovations, these efforts are often limited in scope and primarily focused on aerospace applications. More importantly, these efforts fail to prioritize the establishment of a true digital thread for AFP. This digital thread is essential for overcoming many of the challenges we will discuss. Future AFP platforms, that we will refer to as AFP+, should from the outset seamlessly integrate a digital thread. This involves establishing robust physical, data, and cyber infrastructures. Within the cyber infrastructure domain, AFP+ should explore advanced sensing techniques to enable machine intelligence and perception. By integrating perception, decision-making, and responsive action within the complex AFP process, we will be well-equipped to accomplish the research challenges we are listing below.

8.4.2.1
Material Independent Automated Process Characterization
Following the acquisition of a new AFP machine, the subsequent process characterization and optimization of processing parameters remain heavily reliant on manual trial-and-error and are highly material-dependent. This significantly hinders both the operational efficiency and the functional versatility of the AFP machine. This limitation is frequently encountered when an AFP machine is solely dedicated to the repetitive production of a single composite part, lacking the adaptability to readily accommodate new geometries or material combinations. AFP+ machines should possess the inherent capability to efficiently characterize process parameters across a wide range of material systems. This intrinsic functionality would streamline the manufacturing of composite materials. Firstly, machines should automate the characterization process for all material systems, including thermosets, thermoplastics, and dry fibers. By systematically varying key process parameters such as heating, speed, compaction force, and tow tension, and correlating them with defects, optimal process parameters can be established. This data, combined with pre-existing knowledge of industry-standard quality targets, anomaly detection, and manufacturing quality management, would enable, as a first stage, the single material system automated process character-ization. Furthermore, AFP+ machines should possess complete material independence. This requires generic hardware adaptations that eliminate the need for material-specific modifications (such as heating mechanisms). Combined with advanced methods for material charac-terization and seamless communication with manufacturing equipment, this allows for efficient transitions between different material systems. Ideally, AFP+ would incorporate adaptable hardware, optimized for various material types, enabling inherent flexibility in processing diverse materials. This first function of material independent automated process characterization would dramatically reduce machine setup time, saving valuable time and resources.

8.4.2.2
Automated Process Placement
The increasing complexity of the composites design space, coupled with the ability of AFP to manufacture intricate parts, necessitates a new

generation of design and process planning tools with adaptable path trajectory planning for intelligently adjusting and connecting various fiber layup lifecycles (such as those that we described in Section 3). A key enhancement for AFP+ would optimize layering strategies to balance competing factors like coverage and wrinkle (or other defects) formation. This becomes more particularly true in cases of complex curvature. Traditional planning often overlooks the potential of toolpath optimization, neglecting process and efficiency improvements. Significant opportunities exist to transition from ply-based to course-based (even tow based in certain cases), where courses are laid down across multiple plies in a deconstructed sequence, maximizing efficiency and enhancing performance to achieve AFP+ desired results. This would ultimately provide new and innovative ways to holistically manage the formation of defects and further refine the AFP process.

8.4.2.3
Manual Inspection Annulment for Automated Fiber Placement

AFP inspection phase remains a significant bottleneck despite advancements in process stability. This labor-intensive step, often comparable in duration to the production process itself, is driven by the need to identify and assess potential defects that could compromise structural integrity. The lack of significant progress in in-situ inspection techniques, which remain largely manual, hinders the overall efficiency and wider adoption of AFP technology. It's worth noting that there have been significant recent advancements in this area. However, it's not yet fully integrated or functional across all material systems and still requires substantial fine-tuning. The current gap between advanced AFP technology and outdated inspection methods requires the creation of new, automated inspection solutions. This would streamline workflows and fully leverage AFP+ by integrating automated inspection processing that works across various materials and manufacturing techniques. An AFP+ in-situ inspection system can issue repair directives in real-time, eliminating the need for time-consuming offline inspection. Building on recent advances in machine learning and in-situ inspection, this system would improve the efficiency and reliability of AFP manufacturing by reducing the time between initial processing and final curing, which is crucial for materials with limited shelf life. The AFP+ heads will be equipped with ports and hardware locations to allow for the integration of various visual and sensor technologies to support this in-situ inspection approach. Further development could enable on-the-fly repairs via an integrated robotic arm.

8.4.2.4
Out of Autoclave AFP Composites Manufacturing

The reliance on post-manufacturing processes such as autoclave curing introduces significant limitations to the advancement of AFP technology. Current configurations primarily utilize thermoset prepregs cured in an autoclave or thermoplastic/dry fiber placement followed by resin transfer molding (RTM) processes. However, to fully realize the potential of AFP, it is essential to expand the process chain beyond these conventional methods. Incorporating techniques like overmolding, thermo-mechanical processes such as thermoforming, and wet resin compression into the AFP process chain can unlock new possibilities. This approach seeks to identify alternative process combinations that capitalize on the high accuracy of AFP while integrating faster curing, consolidation, or infusion methods. This would shift the focus from AFP as solely a net-shape manufacturing process towards its potential for creating intermediate shapes. These shapes can undergo further processing and evaluation as either final components or near-net shapes requiring minimal finishing. This approach would expand the versatility of AFP technology, enabling its integration with other manufacturing processes to achieve complex geometries and high-quality composite parts.

8.4.2.5
Conclusion

Identifying necessary advancements to enhance process accessibility, affordability, capability, and integration of recent information technology benefits is challenging. This section focused on principal achievable advancements with significant potential to advance the field. Future editions should assess the proposed functions, evaluating progress and identifying new opportunities. With focused effort and open access, AFP has the potential to generate unprecedented advancements across various domains and truly become a self-organizing production system.

8.5.
Conclusion

This chapter has presented a vision for future manufacturing, as well as the foundation for a new cycle in the realm of AFP. We discussed a digital AFP workflow that integrates design, process planning, manufacturing, and inspection. Following, we presented the importance of hybridizing physics and data-based models for a more accurate prediction. Finally, the vision for an industry of the future cell constructed at the University of South Carolina (Future Factories Laboratory) is presented to connect the aforementioned concepts. In conclusion, scientific efforts are geared toward enacting and defining the future of manufacturing. This chapter sheds light on the importance of how robotics and automation, cyber-infrastructures, data-based models, usage of computing, and AI can help us define the neXt advanced composite structures. We have concluded this chapter and this book by presenting the key functionalities of future AFP systems, an advancement we have termed AFP+.

References

8.1. Harik, R., "neXt Automated Fiber Placement: Advancing Composites Manufacturing towards a New Paradigm," *SAMPE Journal* (2020): 6-14.

8.2. Greenberg, B., "Design and Control of an Arrayed Infrared (MAT-IR) Heater for Accurate Heating Control during Automated Fiber Placement," Master's thesis, University of South Carolina, 2020.

8.3. Harik, R.S. and Godbold, D., *Towards Smart Automated Fiber Placement*, (Manuscript accepted for Composites Handbook 2026), ASM International, 2026.

8.4. Mtorres, "Digitization," https://mtorres.es/en/digitization.

8.5. Broetje-Automation, "SOUL RMOS Digital Twin & Machine Simulation," https://broetje-automation.de/products/smart-process/digital-solutions/product/soul-rmos/.

8.6. Mtorres, "TorresFactory 4.0, Solution for Your Digitalization Needs," https://mtorres.es/en/news/other-projects/torresFactory-4-solution-digitalization.

8.7. Coriolis, "CoDa," https://www.coriolis-composites.com/software-solutions/coda/.

8.8. Coriolis, "CADFiber," https://www.coriolis-composites.com/software-solutions/cadfiber/.

8.9. Coriolis, "CATFiber," https://www.coriolis-composites.com/software-solutions/catfiber/.

8.10. Addcomposites, "AddPath," https://www.addcomposites.com/all-products/addpath.

8.11. Coriolis, "Human Machine Interfaces (HMI)," https://www.coriolis-composites.com/machines/human-machine-interfaces-hmi/.

8.12. Electroimpact, "Electroimpact's Catalog of Advanced Composite Manufacturing Products," https://www.electroimpact.com/Products/composites-manufacturing/technology.

8.13. Ingersoll Machine Tools, "Automated Composite Structure Inspection System," https://en.machinetools.camozzi.com/products/composite-manufacturing/all-products/acsis-.kl.

8.14. Fives, "Composite Optical Automated Surface Tracking," https://www.fives-group.com/high-precision-machines/composites-automated-solutions/forming-inspection/composite-optical-automated-surface-tracking.

Index

About the Authors

Dr. Ramy Harik

Dr. Ramy Harik, a Fulbright Alumni, is the Director of the Clemson Composites Center and a Professor of Automotive Engineering at Clemson University. Ramy holds a bachelor's/master's degree in mechanical engineering, a master's degree in automated manufacturing, and a PhD in industrial/mechanical engineering. His teaching focuses on manufacturing, smart manufac-turing, and composite manufacturing.

Courtesy of Ramy Harik.

Dr. Harik serves as an associate editor for SME Manufacturing Letters and authored the *Introduction to Advanced Manufacturing* textbook published by SAE. He has secured over 15 million USD in funding from NASA, NSF, Boeing, Toray and others. Recognized as one of the top 20 influential professors in Smart Manufacturing by SME's *Smart Manufacturing Magazine* in 2020, he has extensive teaching experience globally and has supervised over 20 graduate students and founded research initiatives. His book *Manufacturing vs Corruption: Who Wins?* won the 2023 Independent Press Award Distinguished Favorite for Social/Political Change books. In June 2024, Dr. Harik received the SC Governor's Award for Excellence in Scientific Awareness, South Carolina's highest honor for promoting science education and supporting the future workforce.

Dr. Alex Brasington

Dr. Alex Brasington is a distinguished material and process engineer in the defense industry, focusing on the research and development of next-generation materials and processes for the production of composite structures. Recognized as an SME 30 under 30 recipient, Dr. Brasington holds a bachelor's degree in civil engineering, a master's degree in aerospace engineering, and a PhD in mechanical engineering. His expertise lies in composite materials and advanced manufacturing

Courtesy of Alex Brasington.

techniques, particularly automated fiber placement (AFP). Dr. Brasington has in-depth knowledge of AFP path planning and process optimization. He has demonstrated significant leadership in his field as an ACM6 scientific chair, guest editor for Manufacturing Letters, and collaborator with industry-leading companies such as NASA, Boeing, and Lockheed Martin. In addition to professional accomplishments, he is dedicated to developing the next generation of the workforce by assisting in students' research and learning efforts.

www.ingramcontent.com/pod-product-compliance
Lightning Source LLC
Chambersburg PA
CBHW041159220326
41597CB00001BA/2